Architectural Col
in the Professional Palette

Architectural Colour
in the Professional Palette

Fiona McLachlan

Routledge
Taylor & Francis Group
LONDON AND NEW YORK

First published 2012
by Routledge
2 Park Square, Milton Park, Abingdon, Oxon OX14 4RN

Simultaneously published in the USA and Canada
by Routledge
711 Third Avenue, New York, NY 10017

Routledge is an imprint of the Taylor & Francis Group, an informa business

© 2012 Fiona McLachlan

The right of Fiona McLachlan to be identified as author of this work has been asserted by her in accordance with sections 77 and 78 of the Copyright, Designs and Patents Act 1988.

All rights reserved. No part of this book may be reprinted or reproduced or utilised in any form or by any electronic, mechanical, or other means, now known or hereafter invented, including photocopying and recording, or in any information storage or retrieval system, without permission in writing from the publishers.

Every effort has been made to contact and acknowledge copyright owners. The publishers would be grateful to hear from any copyright holder who is not acknowledged here and will undertake to rectify any errors or omissions in future printings or editions of the book.

Trademark notice: Product or corporate names may be trademarks or registered trademarks, and are used only for identification and explanation without intent to infringe.

British Library Cataloguing in Publication Data
A catalogue record for this book is available from the British Library

Library of Congress Cataloging in Publication Data
McLachlan, Fiona, 1958–
Architectural colour in the professional palette / Fiona McLachlan. — 1st ed.
   p. cm.
Includes bibliographical references and index.
1. Color in architecture. 2. Architecture, Modern--20th century—Themes, motives.
3. Architecture, Modern—21st century—Themes, motives. I. Title.
   NA2795.M45 2012
   729--dc23
   2011040095

ISBN: 978-0-415-59708-1 (hbk)
ISBN: 978-0-415-59709-8 (pbk)
ISBN: 978-0-203-12147-4 (ebk)

Designed and typeset by Alex Lazarou
Printed and bound in Great Britain by Ashford Colour Press Ltd, Gosport, Hampshire

– CONTENTS –

Preface, ix

ONE
Introduction: investigations in the professional palette, 1
Colour under threat, 1
Meaning in colour, 2
The power of colour, 3
Investigations in the professional palette, 4

TWO
Colour, form and material surface, 8
Purple shadows, 9
The perception of form, 11
Surface, 14
Material surface: a short history of pigments, 18
Contemporary pigments, 21
Pigments and health, 24
Pigmented materials and the future, 25
Colour, form and material surface, 27

THREE
The unattainable myth of novelty: Caruso St John, 28
Colour and cultural tradition, 29
Victoriana at Bethnal Green, 30
Fashion and anti-fashion, 36
Dressing and wrapping, 40

FOUR
An intuitive palette: O'Donnell + Tuomey, 48
Colour in the ethos of the office, 50
Colour, form and surface, 51
Meaning and association, 55
Seriality in the palette, 57

FIVE
Who's afraid of red, yellow and blue? Erich Wiesner and Otto Steidle, 62
Artist/architects, 63
Heroic Modernism, 64
'Colour Field' and the emancipation of colour in the 1950s and 1960s, 67
Authority, originality and the creative impulse, 69
Wiesner and Steidle: artist/architect collaboration, 71
Rhythm and blues, 72
Experimental echoes, 79

SIX
Place, space, colour and light: Steven Holl, 82
Colour and light, 83
Temporality, 87
Location and translocation, 92
Precision or doubt?, 97
Space, place and time, 101

SEVEN
Surface and edge: Gigon/Guyer, 102
Genius loci, 103
Variety as identity, 107
High gloss: surface reflectance, 110
The power of the edge, 113

EIGHT
Memories, associations and the brightness of yellow: AHMM, 118
Colour in the urban realm, 119
'Signing' rather than signage, 120
Parrots among the pigeons, 123
Developing the professional palette, 128
The brightness of yellow, 130

NINE
Synergies and discords: Sauerbruch Hutton, 136
Between the physical and the visual, 137
2D to 3D, 139
Well-being, 140
Subverting the form, 142
Irregularity in the facade, 145
Harmony and dissonance, 149
Equilibrium, 155

TEN
Transformational, instrumental colour: UN Studio, 158
Amorphous space, 159
The instrumentality of colour, 162
Experimentation, 164
Illusion, 165
Transformational colour, 170

ELEVEN
Navigation, communication and language, 176
Navigation systems, 177
The art of communication, 181
Digital colour space, 182

TWELVE
Playing space: laws, rules and prescription, 184
The *wrong* blue: subjective experience and objective recognition, 185
Laws, rules and prescription: didactic approaches, 187
Colour choice and composition, 191
Between pragmatism and the sublime, 193
Conclusion, 194

Acknowledgements, 196
Notes, 197
Bibliography, 219
Image credits, 227
Index, 229

Dulux
colour mixing
FOR WOOD & METAL
62
63
44
54
23
34
24
35
55
56
25
36
48
49
26
37
16
27
38
17
28
18
29
8
9
19
30
10

PREFACE

*'[T]his book does not follow an academic conception of "theory and practice".
It reverses this order and places practice before theory, which,
after all, is the conclusion of practice.'* [1]

My aim in writing this book is to consider the application of colour in professional architectural practice by reflecting on the work of a selected group of contemporary architects who use colour in a confident manner.[2] Much has been written on colour theory and science, but little on the everyday decisions made by architects who may, sheepishly, admit that their use of colour is uninformed and somewhat arbitrary. The book is structured as a series of thematic essays based on interviews with contemporary architects, teasing out philosophical, practical, physical and aesthetic themes within their approaches, using examples drawn from their architecture and from their practice. While there is a given order to the chapters, some readers may prefer to dip into the essays, out of sequence. The book is not intended as a manual for colour use, or to dictate or prescribe a particular methodology. Most architects would strongly resist the idea that the use of colour in architecture can be systemized, or separated from a holistic approach to architectural design. Instead, the book uses the work of architectural practices to introduce aspects of colour theory, in order to provide insights into how specific contemporary architects use colour as an integral part of their design ethos, and as part of the palette of materials and design tools available to them. The text is intended to be stimulating to and directly influential on practising architects in a pragmatic manner, as well as inspiring students and professionals. As William Braham has suggested, 'what architects really require is a logic of the color concepts that influence and organize their work rather than any unified theory of color physics, perception or psychology'.[3]

facing page
Paint pots numbered in series for triptych painting
Investigations in the Professional Palette, see page 6

For my mother
Dr Audrey Miller

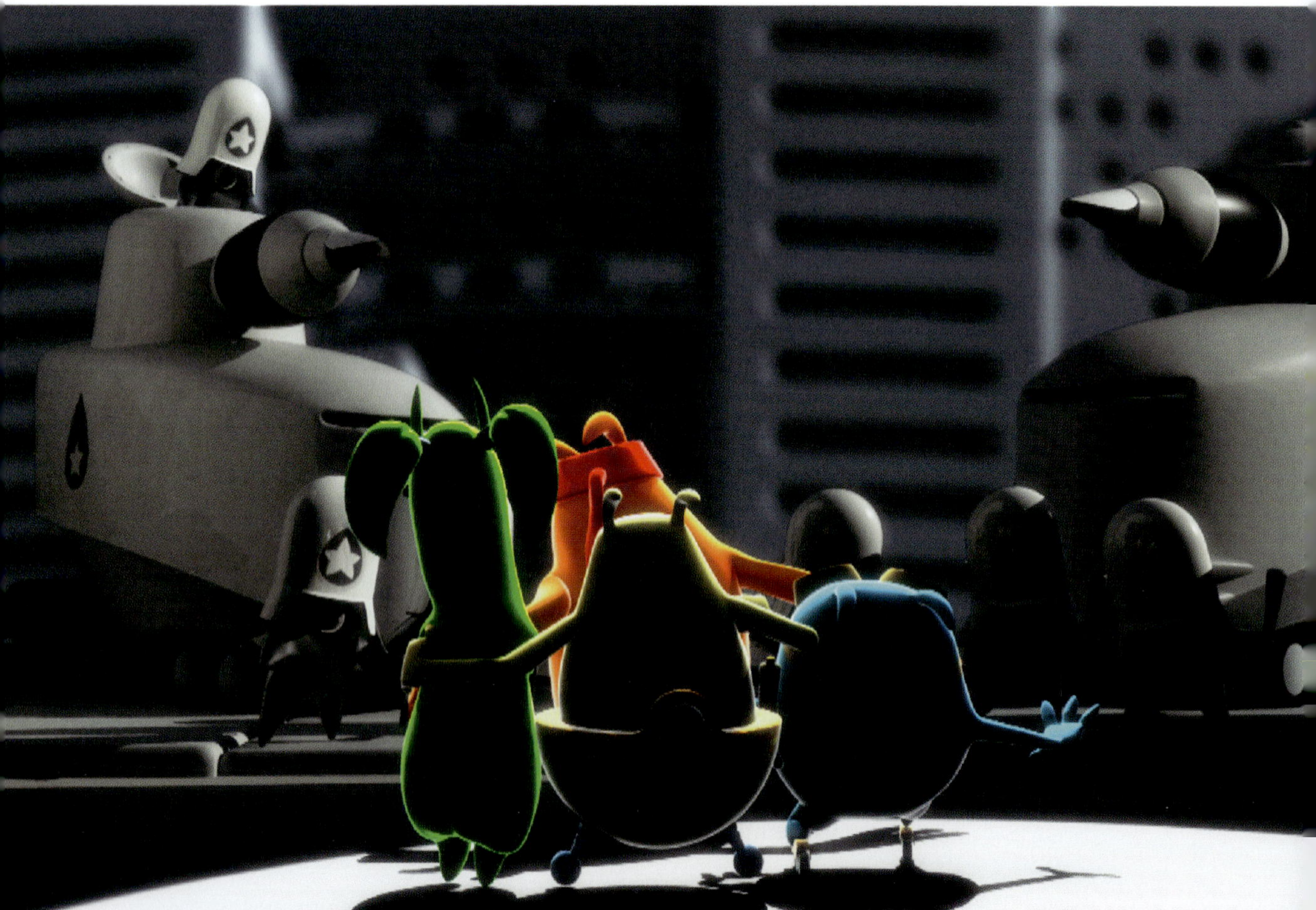

– 1 –
Introduction:
investigations in the professional palette

Colour under threat

THE artist Eugène Delacroix criticized the French art academies in the early nineteenth century for not teaching colour:

> The elements of colour theory have neither been analysed nor taught in our schools of art, because in France it is considered superfluous to study the laws of colour, according to the saying 'Draughtsmen may be made, but colorists are born'.[1]

Two centuries later, it seems that a similar attitude to colour persists. Colour is seen as dangerous territory. David Batchelor's book *Chromaphobia* eloquently explores what he perceives as a general state of anxiety, or wariness, in relation to the use of colour in art and architecture. A computer game 'de Blob' (2008), portrays colours as heroic anarchists defying an invasive monochrome authority, which has literally sucked the colour out of the town with a giant vacuum cleaner.[2] The anti-establishment blobs bounce around on the surfaces of buildings, instantly transforming the city and invigorating its repressed citizens. An underlying message is evident – colour is joyful and represents freedom and individual expression. The dull grey authority can also be interpreted as an oppressive, over-controlling hand, curtailing or schooling the freely expressive child in the expectations of the adult world. Strong colour has long been characterized as naïve, lacking in sophistication, literally childish, whereas a muted palette, or monochrome, by contrast, denotes refinement. In Western society, colour has been associated with femininity, and with emotional, rather than rational or logical responses, although there is little physiological evidence of any true gender difference on which to base such an assertion.[3]

facing page
Still image taken from 'de Blob' video game,
Blue Tongue Entertainment, Australia (2008)

The general nervousness about discussing colour has led to a situation where it is, to some extent, ignored in architectural education. Students are generally encouraged to model in grey or white card and cast in wax, clay or plaster, encouraging the exploration of form and spatial relationships without reference to colour. Attempts at colour representation are frequently garish, clumsy and, at the scale of a drawing, can be overly dominant, and so are often best avoided. Colour does appear, but more commonly as inherent in materials: timber, steel, stone, brick and glass. In the early 1980s, coloured hand drawings, exemplified by the sketches of Aldo Rossi, Cesar Pelli, James Stirling or Nigel Coates, were the norm, bringing the conceptual use of colour into both design process and representation. Perhaps the advent of crude, computer-generated surface rendering has been partly to blame for this exclusion, but the problem seems deeper than one purely of abstraction or representation. Colour theory is taught in a few schools, yet many architects will admit to a lack of confidence with colour, or at least to a sense of arbitrariness in choice and application. Many practitioners feel ill-equipped to explore the potential of simple pigments to alter, tune or define the character of a space or surface. Similarly, clients tend to associate an understanding of colour with interior designers, rather than architects. Could this lack of confidence relate to the inability to exert complete control over the effects of such a relational medium as colour, or is it simply driven by a state of ignorance?

Meaning in colour

'What if Darth Vader wore pink?' [4]

Colour has associations, in terms of social and cultural history, in relation to its application and in multiple interpretations of its meaning. It is impossible to conceive of Darth Vader in pink as being remotely intimidating. Indeed, the very image, however innocently noted, is strongly subversive, not just because of the relationship between complete darkness and evil, but also that between the colour pink and its association with clichéd girlie attributes and gay culture.[5] The meaning of specific colours varies in different societies and cultures and is therefore not constant. An advertising campaign for an international bank highlighted the importance of local knowledge in relation to the meaning of gestures.[6] The same is true of the meaning of colour. In Thailand, for example, each day of the week has an ascribed colour, and people wear the colour of their birthday for special occasions. In the built environment, colour can be inappropriately used if its local meaning is not understood; for example, territorial football allegiances can prohibit or add significance to the use of certain colours. The Allianz stadium in Munich, shared between two teams, is illuminated with blue or red light to identify which team is playing. The international meanings, associations and nomenclatures for colour have been widely studied and so do not need rehearsing here. Colour has been associated with shape, with emotional responses, and with personality traits. People may live with self-imposed dress restrictions, believing their outlook, mood, performance to be dependent on colour, while for others, colour is simply cosmetic and inconsequential.

The power of colour

In the 1950s and 1960s, the American 'Colour Field' painters, such as Barnett Newman, Jackson Pollock, Mark Rothko and Robert Motherwell, re-awakened interest in the Sublime. In the eighteenth century, the awe-inspiring power of nature, either physical or metaphysical, had been a strong influence on art and literature, inspired by Edmund Burke's *A Philosophical Enquiry into the Origin of Our Ideas of the Sublime and Beautiful*.[7] The large canvases of Rothko and Motherwell invited the viewer to be enveloped in an ecstatic void of colour. The scale was critical. They rejected European representation in favour of expressions of the vastness of the American landscape. The paintings are filled with colour, yet empty of form. They focus on the moment of encounter, of here-and-now. Barnett Newman announced 'we are re-asserting man's natural desire for the exalted, for a concern with our relationship to the absolute emotions'.[8] Emotion and colour are therefore intertwined. Colour can exert a powerful tug on our state of mind, on our well-being. The colour light installations by James Turrell induce an overwhelming response. The intense colour, and apparently

James Turrell: Bridget's Bardo,
The Wolfsburg Project, Kunstmuseum Wolfsburg (2010)

limitless space are sublime experiences, and are simultaneously puzzling, calming, invigorating, unfathomable – but always experienced from a place of safety. This is art, after all; one can walk away. Not so with architecture. The scale of architecture suggests that it would be overpowering, and therefore irresponsible, to unleash the metaphysical power of colour, as experienced in a Turrell installation on everyday activities, but it is vital that we do not underestimate the potential strength of this tool as an integral component of architectural design. The work of architect Steven Holl exemplifies the subtle control of colour and light, to stir rather than terrify, to lift the spirit, and to engage body and mind in space.

Investigations in the professional palette

THE starting point for this book was a project undertaken to document the use of colour by my own architectural practice over nearly 30 years.[9] The colours used had been originally selected for various built works, predominately on the basis of instinct and context, without a systemized or theoretical basis.

The catalogue to the 1994 exhibition *Tales from Two Cities*, curated by Shane O'Toole, documented the work of emerging architects in Edinburgh and Dublin, and identified consistent devices in the work of each practice.[10] At that time, most of my practice's work comprised small-scale insertions and re-modelling of existing, sometimes historically notable, buildings in and around Edinburgh. The historic city core, a World Heritage site, has minimal applied colour and a colour palette narrowly restricted by planning constraints. Similarly, the Georgian and Victorian developments are characterized by the structural polychromy of creamy yellow sandstone, blackened with time and soot, white-painted window frames, blue-grey slates and little else. Such is the control over applied or artificial colour, that window frames in the areas of the New Town area of Edinburgh (1789–1820) must be white, not brilliant white, and railings should be black.[11] One will readily accept, however, that in such a strong urban setting, the visual unity of the communal superblock should have priority over any individual expression by the owners of each dwelling – externally at least. Our approach to projects in this context was to heighten the individuality of the interior spaces, by remodelling walls, floor and ceilings to redefine the form, division and proportions of each room. New architectural elements were inserted as if they were pieces of large-scale furniture designed to suit the new usage. The form of these insertions made a number of adjustments to the use and perception of the rooms. Elements were set to constrict or open up space, to hint at rooms within rooms, orientate towards a particular view, provide enclosure to a specified function, give focus where needed or clarify a threshold between public and private usage. Ceilings responded to the plans, reinforcing the volumes and forms.

Colour was also an integral part of this process. The aim was to respect the original features while redefining the rooms by inserting highly coloured, crisply detailed modern elements. Colour was used to contrast old against new, exuberance against simplicity. The underlying principle with respect to the use of colour was to retain the traditional architecture of the original shell, while

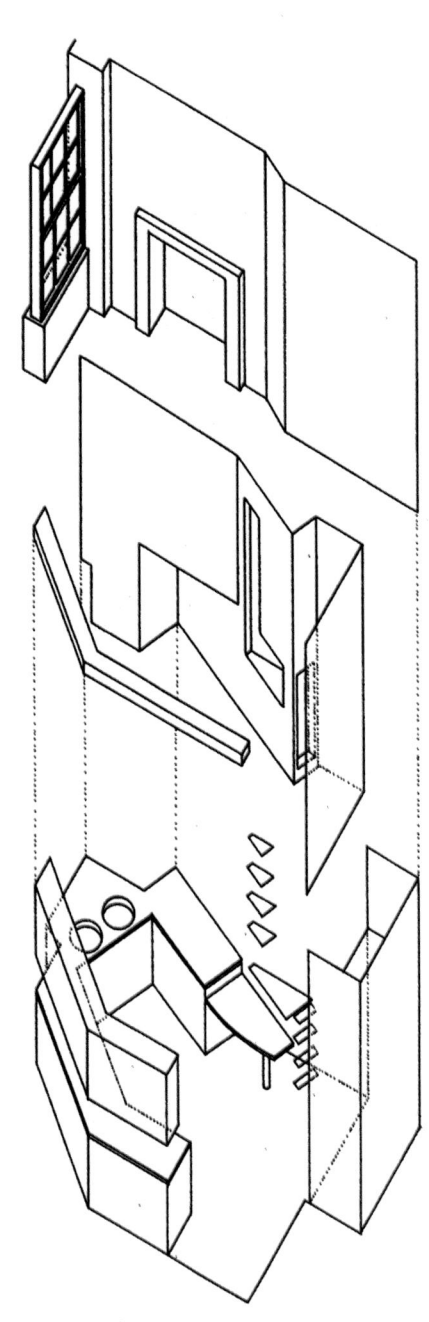

Elements of form: abstraction of architectural elements for flat alterations, Edinburgh, E & F McLachlan Architects (1992)

muting it to allow the modern insertions to assert themselves through their more aggressive colours. For the *Tales from Two Cities* exhibition, the curator asked for a series of monochrome drawings and photographs to disassociate form and colour. The colour was thought to mask the form, yet the neo-plastic formal elements, that appear to thread through walls, were defined by their colour to support the illusion of dissolution of form. The colour, in this case, is less about a surface application and more a way of ordering spatial hierarchies.

The actual choice of colours was secondary to this conceptual position and was made on an empirical rather than theoretical basis. Yet while there was no overt theoretical position on colour, it became evident in retrospect that there were tacit guiding principles. Many architects, it would seem, operate in a similar manner. They are unlikely to follow the teachings of the colour theorists such as Johann Wolfgang von Goethe, Faber Birren, Josef Albers or Johannes Itten, but neither is their selection of colour completely random. To what extent do subjective choices overrule or influence objective principles in colour design?

Johannes Itten's experiments with students at the Weimar Bauhaus, documented later in *Kunst der Farbe [The Art of Color]*, aimed to reveal inherent responses to certain colours.[12] It was thought that gaining an understanding of each individual's 'subjective timbre' would motivate and guide the student. Both Itten and his successor at the Bauhaus, Josef Albers, used practical experiments to document such subjectivity, and there have been many subsequent studies documenting colour preference across different cultural and social groupings, and by gender.[13] Itten and Albers also used practical experiments to teach objective principles of colour theory, such as the interaction of one colour with another, the relative intensity of hues, and simultaneous and successive contrast, which were explored through the use of coloured paper collage. They

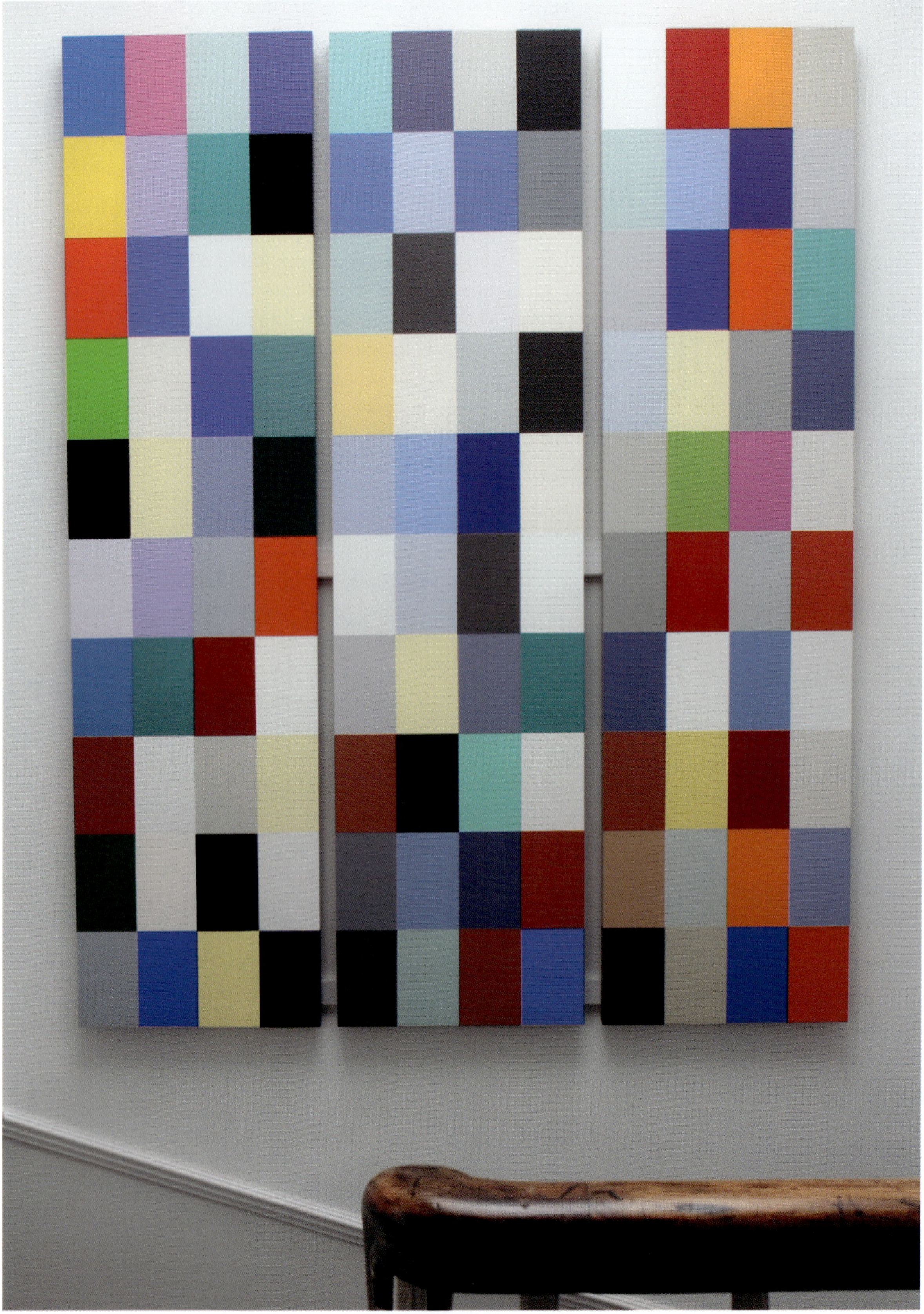

drew on nineteenth-century texts by Goethe, Philipp Otto Runge and Michel Eugène Chevreul, which sought to document objective colour effects and give guidance on the application of colour.

Architects may also be unaware of their own subjective preferences, or recurring combinations of colour, in terms of a particular 'palette' in their work. This was the case with my practice, and was revealed by a research project that preceded this book. Initially, the research centred on establishing a 'chronology of colour', by poring over old job files for decoration schedules. Aided by digital software, the chronology was indexed as swatches.[14] The output was in the form of a triptych painting that represents the colours specified over a 30-year period: 1980–2008.

The composition of the painting is deliberately abstract, yet behind each subset of colours is a narrative related to each project, not readily evident from the painting, but known to the architects. The abstraction was essential in order to isolate the colour from its context; from the influence of form, surface, material, light, function, meaning and atmosphere of the space for which it was originally conceived. Each colour was constrained and rationalized by a grid to disassociate the actual colour still further and perceive it as close as possible to pure colour, equalized in scale but not in intensity. By placing the colours contiguously, rather than with a border, there is visual interaction that affects the perception of individual colours. Boundaries between opposing colours in the spectrum appear to have harder edges than the muted boundaries between adjacent hues. Edges between these are softer and ambiguous, and the colours appear unstable, de-materialized. Undoubtedly, there is subjectivity in the palette, but this is initially tempered by the typology, with softer hues used for social housing, strong theatrical colours for public spaces and more nuanced colours in projects for private clients' own homes. Furthermore, even in this very small study the influence of the period in which the colours were specified is clearly evident, with primary colours only used in the 1980s, and strong oranges first appearing in 2006. Opinion of what is good or acceptable in colour choice changes with time. This is not merely a product of fashion, which will be explored further in Chapter 3, but reflects the unstable nature of public opinion and perception. There are predominant colours in the palette, a recurring pale yellow, which was unexpected, and a temper leaning towards warm, purply-blues, blue greens and brick reds.[15]

The composition was made after a number of iterations. Pure indexing proved unsatisfactory as an ordering device, as the triptych became a physical object, had its own context and needed to be composed. Although the final arrangement acts within self-imposed rules of adjacency, chronologically from left to right, and with projects related vertically, the final configuration was adjusted by eye.[16]

Although abstract, the process of making the painting revealed a number of factors that were at play in the ethos of the architectural practice, with the colours themselves, the interaction between them, and the conceptual framework, all of which are further explored in the following chapters through the work of eight contemporary architectural practices. Each chapter focuses on the most significant underlying principles in the use of colour by the architects, some of which are idiosyncratic of each practice and their architecture.

facing page

Investigations in the Professional Palette,
Fiona McLachlan (2008)

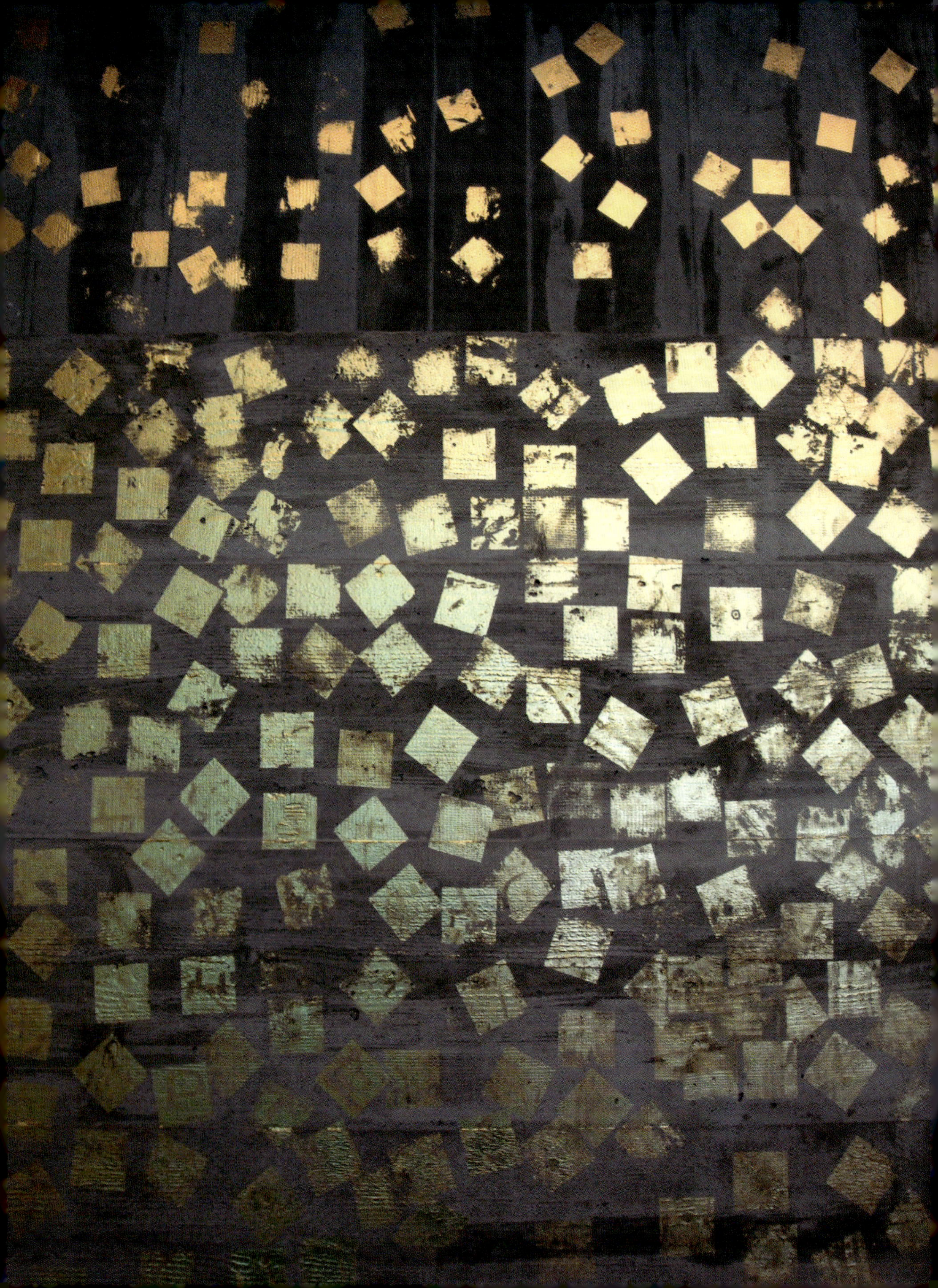

– 2 –
Colour, form and material surface

'I know that my colours are not yours. Two colours are never the same, even if they are from the same tube. Context changes the way we perceive them ... I've placed no colour photos in this book, as that would be a futile attempt to imprison them ... I prefer that the colours should float and take flight in your minds.' [1]

Purple shadows

COLOUR is a complex phenomenon – unstable, unreliable, subjective, relational and entirely contingent on time, location, form and material surface. It can provoke a sense of unease among designers and architectural commentators, particularly if considered as dissociated from its context. In architecture, beyond the inherent natural colours of material substance, it is most frequently introduced in the form of a layer of pigment applied to surfaces, or mixed into man-made materials as a tint.[2] A thin skin of paint will protect and modify appearance. It may be as simple and deliberate as whitewash, it may be loud and strong or soft and subtle – indeed colour may not be inherently colourful. It represents choice, a further decision among the thousands of decisions that constitute a conventional design process. Many such decisions are logical, rational and reasoned, and so are easily justified. Colour, by contrast, represents the irrational, the emotional. To some designers, the inability to exercise precise control over its impact threatens their sense of authority.

Aristotle, Isaac Newton and Johann Wolfgang von Goethe are among an array of scholars who have gradually edged humanity towards an understanding of the complex phenomena associated with colour, but it is only in the past two centuries that the relationship of colour to material surface has been more fully understood. Originally, it was thought that colour was intrinsic to the object, whereas it has been clearly established that colour is generated by light being reflected off the

facing page
Gold leaf on in-situ concrete at the Rietberg Museum, Zurich, Adolf Krischanitz Architects (2007)

object's surface.[3] Colour registered by the brain is generated by light from the part of the spectrum that is not wholly absorbed by the object. A 'blue' pigment applied to a surface makes it appear blue because the chemical composition of the pigment absorbs all light hues except blue, which is reflected back from the surface and observed by the eye.[4] The colour we see is therefore dependent on the physical properties and chemical compound of the surface. It can be hard for the mind to accept when staring into a pot of paint that this liquid colour is entirely a product of the reflected light, or that such a simple product can generate powerful experiential phenomena. Goethe was unable to accept that coloured light when mixed produces white light, yet pigments when mixed produce a muddy grey darkness. He was roundly ridiculed at the time for daring to criticize such an eminent scientist as Sir Isaac Newton. Once it is understood that it is light that generates colour, it is clear that colour must be a contingent phenomenon, as light is not a stable condition. Johannes Itten, one of the key figures in twentieth-century colour theory, makes reference to the optical effects of colour as part of an essential understanding of colour perception.

The Impressionist artists, notably Claude Monet and Vincent van Gogh, are known to have read Michel Eugène Chevreul's seminal text *The Laws of Contrast of Colour: and Their Application to the Arts of Painting, Decoration of Buildings, Mosaic Work, Tapestry and Carpet Weaving, Calico Printing, Dress, Paper Staining, Printing, Illumination, Landscape and Flower Gardening, etc.* which was published in France in 1839.[5] He defined the effects of *simultaneous contrast*, noting that hues, located as opposites on his colour wheel, for example blue and orange, will appear to strengthen each other when placed adjacently. Chevreul's aim was to write in such a way that the theories were presented practically and in relation to numerous quotidian applications. The Impressionists caused considerable consternation by painting shadows of sunny landscape scenes in purply-blues rather than in grey-black, which was considered the correct approach at the time. This heightened the effect of both the shadow and the subject. Chevreul was drawing from Goethe's earlier text *Farbenlehre (Doctrine of Colours)*, which suggested that tinted shadows, particularly lilac shadows on snowy mountains, could be observed if the viewer was open-minded enough to let their eyes overrule the expectation of the mind.[6] Monet was said to have proclaimed 'I have finally discovered the true colour of the atmosphere. It's violet. Fresh air is violet. Three years from now everyone will work in violet.'[7]

Contrasting colour at
House P, Edinburgh,
E & F McLachlan Architects
(2005)

Artists and scientists are extremely aware of the way in which adjacency of colour influences perception. The concept of the 'after image', or *successive contrast*, established that the eyes generally make an opposite spectral colour, after exposure to a strong hue. Practical studies of physiological effects by Johannes Itten and by Josef Albers (subsequently documented in his *Interaction of Color*), were intended to arm students across a wide range of disciplines, with a grasp of colour theory. They followed similar principles to Chevreul and illustrated that the colours we perceive will be affected by the background or base colour, adjacent colours or surfaces, as well as by light conditions and climate. The effect of colour is also very different externally and internally, not purely because of the light conditions. Externally, the observer is generally remote, and sees the object or surface without necessarily becoming involved. Internally, the effects of colour can be heightened by the immersion of the body in the space, making it less easy to remain detached from the experience. When colour is applied to form or to surface its appearance is modified, not always in a predictable manner.

The perception of form

'*Form is absolute and colour wholly relative.*'[8]

IN consideration of form and surface, it is necessary to be aware of how the mind perceives form. We distinguish an object from its context by reading objects from the edges inwards. Similarly, with text, the mind scans words from the outside inwards, constructing the words from our expectations of what should be there. Take, for instance, this example used by the popular stage mind-reader Banacheck:

> it deosn't mttaer in what order the ltteers in a word are, the only iprmoetnt thing is that the frist and lsat ltteers be in the rghit pclae. Thhe rset can be a total mses and you can still raed it wouthit porbelm. Tihs is bcuseae the human mind deos not raed each lteter by istlef, but the word as a whole.[9]

Camouflage in birds and insects works in a similar manner. While moths that are a single colour stand out against a background, and multi-coloured or patterned ones merge more effectively with their surroundings, those in which the pattern extends to the edges and breaks the line of the edge are the most effective at dissolving into their environment.[10] The effect has been used as disruptive camouflage on warships to distort and break up the apparent scale of the object. The artist Laurent La Gamba uses the effects of camouflage in his 'homochromie' series. The edges of the human form are broken using acrylic paint on body suits, which mimics the background colours.[11] By exactly matching the pattern to a complex background, the object – in his case the human body – becomes indistinguishable from the background.

If the edges are the key, the use of colour to define specific architectural elements makes sense. The form of the element is heightened and accentuated by the relative difference in colour

Chequered flats at Dalston, London, AHMM (1999)

between the object and its surroundings. Choosing a colour similar in tone to the context will reduce the impact of an object by lessening the contrast. Using colour to deliberately break up the edges of a room, crossing or wrapping around corners, will have the effect of making a more distorted perception of the space. Colour is more immediately perceived than form, and will tend to dominate the reading of an object. Architectural form therefore, like natural form, can be made to be more or less conspicuous. Colour may either reinforce the form, or camouflage through a distortion of the perception of the volume, or the surface. For example, architects Sauerbruch Hutton use polychromatic surfaces that wrap architectural forms and subvert surfaces. They note 'the optical dissolution or manipulation of form introduces uncertainty, doubt, ambiguity. It engages you.'[12]

The development of the structural frame, dissociating enclosing wall elements from structural requirements, demanded a reconsideration of the use of colour. The subsequent development of the free plan meant that the use of colour was liberated from its traditional use on the enclosing walls and ceilings of conventional rooms. Le Corbusier searched for a new logic to respond to the freedom of flowing, three-dimensional space. Colour was used to highlight specific elements of form as an organizing device, or rarifier, within a unified conceptual hierarchy, in order to create 'colour solidity – or the illusion of solidity'.[13] In Villa La Roche (1923), for example, the chimney is highlighted by colour, as is the curving ramp in the study, expressing them as elements of form within the overall composition. Le Corbusier was initially uncomfortable with using colour in an abstract manner to imply the dissolution of space or induce optical illusions, as practised by the artist Theo van Doesburg and the de Stijl architects. In Gerrit Reitveld's *Schröder House*, Utrecht (1925), formal elements were decomposed into planes of colour, as in two-dimensional painting.[14] Colour was used as an element of spatial construction, not merely as a decorative coating. The colour fields effectively hang in space, undermining the unity and permitting the space to be perceived as free-flowing, emphasized by the corner windows that merge interior and exterior space. More recently, the artist Oscar Putz experimented with polychromatic painted panels in the Kix Bar in Vienna (1986). Colours wrap around the interior surfaces, but also cross the edges of the form. What is perceived is a psychic effect opposed to the physical fact – the application of the colour to surfaces disrupts the spatial reading of the interior volume of the bar.

As with any other skill, once the principles are mastered, the architect can begin to challenge, radicalize and question. Colour can be used to compose an intellectual illusion in support of a particular reading of form and surface. Exterior forms that appear solid may be incised to suggest colour within, like slicing into a fruit, either by indenting the form at a reveal, as in O'Donnell + Tuomey's Housing in Galbally, Ireland (2002), or using twisting geometry and cuts to expose coloured sinews, such as in UN Studio's Research Laboratories at Gröningen, Netherlands (2008). Colour may support a codification of physical expression, for example at Peter Eisenmann's Columbus Convention Center (1993), while complex spatial geometries employed by Zaha Hadid and UN Studio suggest a further extension of colour in relation to ambiguous, amorphous form or continuous surface, such as the folded planes of the orange theatre by UN Studio at Lelystad, Netherlands (2007).

Kix Bar, Vienna, Oskar Putz (1986)

Surface

GOTTFRIED Semper (1803–1879) discussed colour in his book *Style in the Technical and Tectonic Arts: Der Stijl* as 'the most subtle and bodiless of covering materials, capable of "masking" the materiality of the stone, so that it appears as pure form'. Semper argued that architectural expression was not limited solely to the representation of construction, as explained by David Leatherbarrow and Mohsen Mostafavi (2002), 'like mimesis in the theatre, which entails the masking of the actor's feelings to make way for the characterization of his role, architectural mimesis masks the materials so that the symbolic content can emerge'. More subtly, they note that cladding or dressing of a surface 'invites the spectator to participate in a tectonic "as if"; that is the building appears as if it were built in other materials, in another time'.[15]

The English Victorian architect Augustus Welby Pugin (1812–52) held firmly to the belief that any surface decoration, such as his many colourful patterns for wallpaper, must emphasize its two-dimensional nature to give emphasis to the solidity, unity and flatness of the wall. He condemned the nineteenth-century fashion for *trompe l'oeil* wall paintings as verging on blasphemous, and argued that the integrity of the wall as solid supportive element was paramount. It should not, therefore, be subject to optical illusion. Contemporary architects, for the most part, have been raised in a Modernist tradition where the spatial effect is considered of primary importance and the surface treatment secondary. Surface has forgotten its significance as the bearer of iconography or of lyrical pattern. In architectural education, one is taught to develop first control of spatial relationships and only then of the structure, construction, environment and materiality. The prevailing attitude is that form should dominate over surface. Surface, for a majority, has been relegated to that which encloses or defines the elements and gives material quality to the sense of space. Surface treatment is considered as secondary, there to enhance the legibility of the primary architectural elements, or the spatial relationships of volume, or to influence the mood or atmosphere of the space. Adolf Loos established the principle that materials should be expressed in their true form, unadorned and not disguised. Leatherbarrow and Mostafavi continue:

> This fact of material formation is also true for a wide range of architectural materials: not only thick and palpable materials, such as stone, timber and metals, but also thin and liquid coverings such as stains and paint. All of the materials of the Looshaus façade – the stucco, marble, steel, glass, and copper – should be interpreted in this way, for each has its own language and method. The application of paint to the surface of another material is also conditioned by this same rule of disallowing the imitation of another material. Accordingly, metal can be painted any colour but a metallic colour.[16]

facing page
Sliced corner at Research Laboratory,
Groningen, Netherlands, UN Studio (2008)

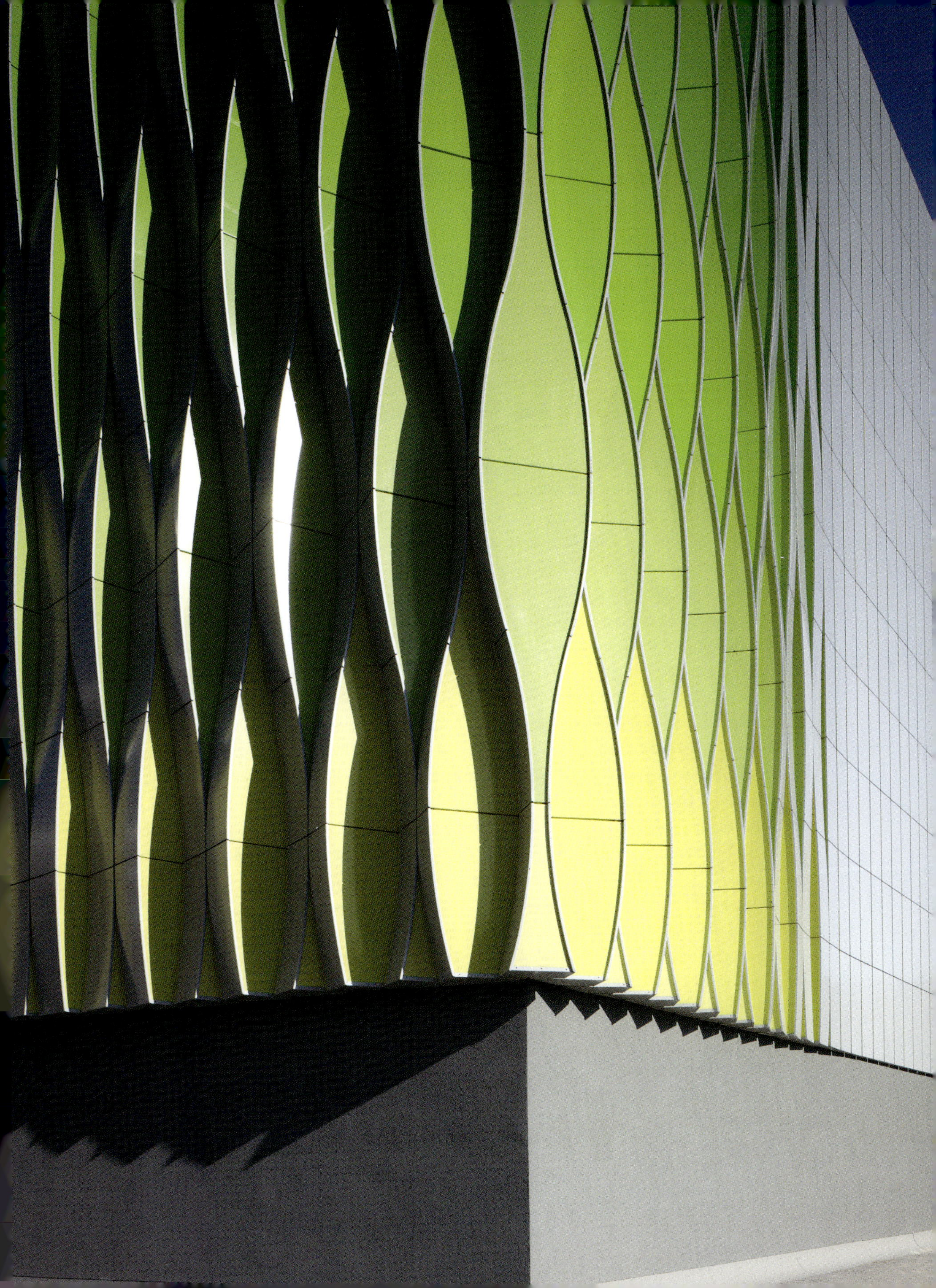

This may be interpreted in two ways; Loos is suggesting that we should either use materials in their natural undecorated state, or, alternatively, painted surfaces are permissible, as long as they clearly express themselves as such, and do not pretend otherwise. For the most part, architecture of the twentieth century adopted the former approach, thus creating an ambiguous relationship between surface decoration and form. Indeed, the architectural pendulum has swung so far that many architects now feel ill-equipped to consider surface decoration of any kind and, most particularly, any form of pattern, narrative, or symbolism. Dismayed by the lack of use and knowledge of symbolic colour and pattern in contemporary architecture, John Outram berates architects who have lost their way in the world of colour. He has been known to post-rationalize – weaving elaborate narratives around decorative elements, such as protective guardian angels who spread their wings over an otherwise prosaic industrial unit. For Outram, it is unforgivable to apply colour as an unthinking, last-minute decision, as is the norm, he argues, in most contemporary work. This issue is further explored in Chapter 8.

The death of ornament, heralded in Loos' *Ornament and Crime*, has been so widespread among architects and architectural educators that the use of pattern is generally left to artists, textile designers and interior designers. Colour, on the other hand, is increasingly present in contemporary architecture and firmly embedded as part of the architectural language of a number of renowned architects. Simpler to conceive of than ornament, it is nevertheless highly conspicuous and potentially gauche, if used unthinkingly or unknowingly.

Mark Rothko's deep red, burnt orange and black abstracts, painted for the Mies van der Rohe Seagram building in 1959, invite the viewer into the implied space beyond the surface of the canvas. Standing momentarily in front of one of his canvases, the effect is simultaneously calming and absorbing. The context of the canvas itself, in relation to the edges, confines and defines the frame. Rothko's later paintings deliberately stop short of the edge, leaving an unpainted border that emphasizes the framing effect for the viewer. The context in which his paintings are exhibited contributes to the viewing, both spatially and, crucially, in terms of the light conditions. In the case of some paintings, such as the 'Cage' paintings by Gerard Richter, the serial nature of the works, which are intended to be read in a particular order, dictates the spatial relationships of one painting to the next, and can cause difficulties for the curator.[17] Richter's directing of the hanging is unusual, as an artist generally has to assume that the context in which their paintings are viewed will vary from exhibition to exhibition. The viewer's interface with the painting as a physical object is through its surface, which, as Richter notes, is the 'key moment of encounter'.[18] The artist Adrian Schiess employs highly reflective surfaces, often laid flat on the floor, to invite changing images, generated by people and the surrounding context. His collaborations with architects Gigon/Guyer make use of similar intensely coloured, shiny panels of glass.[19] The colour is deliberately confused and distorted by the surface reflections.

facing page
Study painting (bottom) and detail (top) of *Virtual Air Conditioners*, Tirana, Albania, Bolles + Wilson (2004)

Colour, form and material surface

Robert Venturi suggested that the Modernist demands for free-flowing space, uninterrupted by the facade, denied the possibility of providing contrast and ambiguity between inside and outside, which he considered an essential characteristic of urban architecture. Venturi argued that the contradictory demands of inside and outside, private and public, should be accommodated within the facade, not necessarily resolved, but expressive of contradiction or discord. 'Since the inside is different from the outside, the wall – the point of change – becomes an architectural event.'[20] The re-emergence of the concept of skin as clothing, as evident in the work of Michael Graves at the Portland Municipal Services Building, Oregon, USA (1982), during the short-lived, pastel-coloured, Postmodernist period, was later considered to be more insubstantial packaging than serious architecture. The recent return of both colour and pattern, for example in the work of Swiss architects Herzog & de Meuron, can be seen as a resurgence of the idea that the building invites the viewer into a dialogue, through the treatment of the surface. The relationship of the surface of an architectural container to that which is contained can also be affected by scale. As Rem Koolhaas has emphasized, the relationship between plan and skin, and even the Modernist concept of plan as generator, ceases to work when buildings are of such enormous scale that the inside can no longer have any direct relation to the outside.[21] At some point in this scaling, the facade becomes a perimeter wall, stemming the flow of the plan. This is particularly true of buildings that have relatively 'dumb' plans, such as sports halls, commercial offices and cultural containers. The surface takes on an independent existence and this may include a narrative of pattern and colour to give meaning and interest.

The architects Julia Bolles and Peter Wilson (Bolles Wilson) have been party to some highly experimental exercises in colouring surfaces. As part of a vast urban painting initiative in Tirana, Albania, by the artist mayor Edi Rama, photos of bland, grey, communist-era buildings were faxed to their office in Münster, Germany. The architects made hand-painted studies and sent the images back to the client, where they were then re-scaled and used to decorate the surfaces of buildings with a thin layer of paint. The result was a complete alteration of the appearance of the city, playful interjections in an otherwise featureless urban landscape. Edi Rama reflects 'it was not an artistic operation … it was just a way to get through this terrible and very, very thick wall between people and authority, and individual life and social life'.[22] Perhaps this act of translation from three dimensions to two, and abstracted from the context, allowed the surface to become more of a simple canvas than a building surface onto which strongly coloured patterns could be applied. The fact that paint can be so readily undone and is easily applied and cheap, makes this less of a risk than more technically sophisticated solutions would be.

Material surface: a short history of pigments

JUST as Le Corbusier observed that the history of the facade in architecture can be aligned with the history of technology, in particular the evolution of the window, so the development of pigments has had a direct effect on painting, textiles, and the use of colour in architecture. This is well-documented territory,[23] but it is useful to highlight some key technical issues pertinent to the contemporary architect. Knowledge of technological innovation, and the availability of pigments in specific forms that can be applied to architectural surfaces, are necessary parts of the process of specification. The practical realities of manufacturing, systems of nomenclature, performance and environmental concerns can, however, make the use of colour complex. The availability and range of pigments and pigmented materials continue to evolve, providing more stable colour in a wider range than has been previously possible. The degree of choice may be seen as either liberating or bewildering.

Earth pigments have been synthesized to provide colour for thousands of years. In antiquity, pigments tended to be used individually and unmixed. Mixing was taboo and was compared to a loss of virginity or purity, as it dulls the vibrancy of the colour. The introduction of oil as a suspension for pigment changed all this, and released a sudden wealth of possibilities using blended colour. As a consequence, natural minerals and other source materials gradually lost their medieval, symbolic virtues. Vermillion, for example, appears less brilliant and green malachite becomes transparent. Ultramarine, seen as the most precious of colours, was observed to turn dark in oil.[24] Sourced in Afghanistan from the mineral lapis lazuli, it was so prohibitively expensive that, along with gold, it was used for only the most significant parts of paintings, with the finest grade of ultramarine being reserved for the mantle of the Madonna. To reduce costs, Florentine painters would substitute the more readily available azurite in place of ultramarine, but this practice was later prohibited by statute. The intense blue vaulted ceiling of the Arena Chapel in Padua, with murals by Giotto from 1305, is an example of an astonishing use of azurite blue.

Interior of the Palazzo Vecchio, Florence (1302), remodelled in the sixteenth century for Duke Cosimo

As a result of this new process, both the meaning and the technical practicalities of painting changed. In architecture, as in painting, pigments had to be mixed fresh, on-site, as they had no 'shelf-life'. Limewashes and distempers mixed with glue were coloured by finely-milled powder pigment. One advantage was that colours could be adjusted to suit the demands of the client and the quality of light. Lime also allows moisture to transfer through the outer surface, which was a critical practical consideration prior to the adoption of damp-proof courses. In the nineteenth century, the range of pigments and dyes expanded exponentially as synthetic products, such as cobalt blue, true orange and violet, were introduced. The Impressionists, in particular, readily adopted the new brilliant colours and, as Philip Ball notes, 'art has never looked back'.[25] In addition to the pigments themselves, technical breakthroughs, such as the advent of paint readily available in portable tubes, which had a longer life, coincided with the growth of landscape painting. Victoria Finlay discusses the gradual dissociation of the artist from a technical understanding of their medium in her book *Colour: A Natural History of the Palette*. When pigments had to be crushed by pestle and mortar and stored in pig-bladder bags ready for use, the artist was immersed in the craft of making the paint and had a better knowledge of how it would perform. By the mid-nineteenth century, paint manufacture in the UK for both artist and house painter had become industrialized and commercialized. Victorians could buy ready-to-use watercolour boxes, effectively establishing art as a hobby. The downside was that artists were vulnerable to fluctuations in the quality of the supplied pigments and there was a growing ignorance of the craft of the paint makers or 'colourmen', who actually prepared the paint. So-called 'fugitive' colours, which subsequently deteriorated badly, flooded the market, radically altering the perceived colour and distorting the impression made by a painting. Indeed, for some, the distortion of age is of positive benefit, as John Constable (1776–1837) observed, 'a good picture, like a good fiddle, should be brown'. The brown tint favoured by Constable was frequently generated by applying a thin film of bitumen to veil the luminosity of the paint. This hue, which was associated with the work of the earlier 'Old Masters', and which seventeenth- and eighteenth-century painters sought to emulate, was actually a result of unstable varnish and had been ridiculed by Hogarth (1697–1764), who was frustrated that these fraudulently dirty pictures sold better than his own.[26]

In *Bright Earth*, Philip Ball notes that the relationship between the availability of pigment and the art created is indisputable, 'with any new invention there are those who sneer that it will corrupt and bemoan the new era'.[27] Ball notes that this was as true of Titian's use of new Venetian colours as it was of the Impressionists' use of chemically-derived pigments. He suggests that the materials have been generally overlooked 'perhaps [as] a consequence of the cultural tendency in the West to separate inspiration from substance'.

By the 1950s, painters had begun to use mass-marketed household paints. Jackson Pollock used enamel gloss paints, while Richard Hamilton used spray paints. More recently, Gary Hume has made use of high-gloss household paint and draught excluder tape for masking. These painters, who work straight from the can or 'off the shelf', willingly have their palette dictated by the manufacturers. Robert Rauschenberg was known to buy cheap, unlabelled cans, and to paint with whatever colours they contained, thus deliberately embracing randomness. Mark Rothko's relationship with paint was at times equally ambivalent. He is reported to have once told a journalist that when the paint ran out, he just bought some more at Woolworths. By contrast, Yves Klein was intensely engaged in the manufacture of the colour, and in the search for a resin that would not dull the intensity of pure pigment. For Klein, the work was the colour and the texture of the pigment, and he was not to be distanced from his materials.

The use of pigments in architecture has seen similar periods of intensity, with peaks of technological innovation, followed by troughs. The century from the early 1800s to the Bauhaus colour courses in the mid-1920s was the most intensely productive period in the advancement of our understanding of colour in both art and science, and was marked by highly colourful architecture. Despite such an explosion in the availability of pigments and pigmented materials for use in buildings, there followed a general reduction in the use of colour on building facades in Western societies, particularly in urban areas, during much of the twentieth century. In the early twenty-first century we are witnessing a further explosion of the potential applications of colour and pattern in architectural design.

Indeed, such is the complexity of contemporary architectural specification, the enormous choice of products available, and the internationalization of manufacture that few architects have the time or inclination to learn about the constituent elements of the products that they specify. Longevity is closely intertwined with performance and maintenance, and the technical properties and components of external colour coatings, which are subject to weather and light, are more likely to be researched than conventional paints for internal use. More normally, the manufacturer's assurance is sought that the product will perform on the required surface in an enduring, low-maintenance manner, while not poisoning the painters or the future users of the building. In this, architects and designers are more aligned to the Victorian hobbyist than the great artist/artisans who were closely involved with making.

Contemporary pigments

So what is blue, green or red now? Gaining an understanding of the technology of paint and pigments used to colour common building materials may help architects to narrow the gap between art and craft. First, a definition:

> a liquid paint is an engineered product made of several different ingredients that mix to create a specific product with its own unique properties. The selection of components … will affect its stability (shelf life), application characteristics, handling, clean-up, disposal and, most importantly, the performance of the product on which it is applied.[28]

Most paints and coatings used in the construction industry are made from a combination of resin (medium or binder), pigments, solvent and extenders. Essentially, there are two types of paint: oil-based, using resins, and water-based. They have a dual use, as decorative and/or protective coatings to surfaces. The resin is the key ingredient, and is tuned to the type of substrate, the location, and required performance. The same substrate may require very different performance levels in terms of durability, flexibility, surface texture, reflectance, method of application, and colour retention, and so will dictate the resin. The particle size and shape affect the reflectance of the surface and the intensity of the colour, while the molecular structure affects its ability to act as a barrier.[29] Most paints for woodwork manufactured before the 1920s were based on linseed oil and would take up to two days to dry. Modern paint dries quickly due to the alkyd used as the resin, but is more brittle and can crack over time.

Contemporary paint manufacture at the Craig & Rose factory, Scotland

Linseed oils are now rarely used, except by ecological paint manufacturers, but may return in some form in the future, as a sustainable resource. As far as the trade is concerned, 'the resin, or binder is the paint'.[30] The solvent simply makes the paint thin enough to flow and eases application. Additives, or extenders, are materials added to the paint, usually in small quantities, to help it dry more quickly, or flow out evenly to remove brush marks and to prevent skin forming in the can. Additives can also be used to make the surface of the paint film more resistant to marking and scratching. Like any recipe, the composition depends on the availability of readily sourced ingredients, and so has distinct regional traditions. In the UK, calcium carbonate (limestone) is common as an extender, in Finland talc is used and, in America, aluminium silicates.

Traditionally, organic pigments were derived from minerals such as iron oxide, yellow ochre, and cobalt. There is a tendency for organic pigments to be used in decorative products, whereas inorganic pigments will more commonly form part of protective products.[31] Le Corbusier's original *Salubra* wallpaper range (1931) was based on traditional, powdered pigments. He noted, 'I kept most of the "noble range": white, ultramarine, blue, English green, yellow ochre, natural sienna earth, a vermillion, a carmine, English red, burnt sienna earth. For each of these tones I researched, from the mural point of view, the most efficient values.' The colours for the second edition were then modified because of the availability of synthetic pigments, but Le Corbusier warned that there are 'tones produced by modern industry that violently shake our nervous system, but fatigue it as quickly'.[32]

Iron oxides are readily available worldwide, hence the predominance of earthy yellows and browns in most regions. They are very versatile, as the chemical structure is changed by heat, resulting in different colours, and they are popular for their warmth of appearance.[33] They give a good density of colour because of their purity, but are still limited in the range of hues available.

Synthetic yellows give a wider range and can be based on an azo dye, such as benzidene yellows, which are available from yellow through to red.[34] They are non-toxic, but can suffer from problems with light-fastness so tend to be more commonly used for interior paints and emulsions. Yellow can be difficult to darken as it can take on an unpleasant greenish hue, whereas creamy and warm yellows are highly popular. By far the most common pigment in use today is white titanium dioxide, a synthetic pigment. It is non-toxic and extremely stable, and forms the base colour of the industry. It has replaced white lead, which is toxic, but lasts a very long time, as the lead reacts with the linseed oil and remains flexible. Women in the eighteenth century risked poisoning when using the so-called

Le Corbusier *Salubra* wallpaper range, second edition (1959)

'Bloom of Youth' on their faces, a product essentially made from white lead.[35] White lead paint is still used where its unique light reflectance and texture are required, but is likely to require special permission, and in consequence can only be used in significant restoration projects. Although it is not toxic, once the pigment has been mixed into the binder it presents a long-term maintenance problem, as any future sanding may create toxic dust.

Black pigment is used directly in paint to adjust the lightness or darkness and is created with carbon blacks or black iron oxide – a synthetic pigment more common in primers and undercoats because of its relatively low tinting strength. Deep and strong reds can be problematical: the paint tends to be more transparent, and may require many more coats. Traditional red iron oxide – a primer and barrier, perhaps most famously used on the Forth Rail Bridge in Scotland – is naturally occurring or can be made synthetically for a brighter, stronger tint.[36] The most commonly used red pigments in decorative paints are toluidine reds (non-toxic azo dyes, bright and clean with a high opacity) and arylamide reds (non-toxic azo dyes, orange red to crimson). Green pigment is generally a phthalo green that is not purely organic, or phthalocyanide green – a blue-green – with good opacity and resistance to solvents.

Irish Film Institute, Dublin, under construction using traditional ochre pigment in render, O'Donnell + Tuomey Architects (1992)

Blue remains the most popular colour in European societies.[37] Blue pigments tend to have very good opacity and so give easy coverage. Phthalocyanide blue is the most common contemporary blue pigment, and varies from reddish-blue to yellowish-green. It has a high tint strength when mixed with white, and good light-fastness. Ultramarine, from the mineral lapis lazuli, is a complex aluminosilicate; its dense and enveloping colour derives from sodium sulphide. It is weak, however, when used as a tint, requiring ten times as much mineral content compared to phthalocyanide blue for an equivalent result.

Understandably, economics drive the industry and almost all pigments now used in building materials are derived from industrially produced inorganic pigments.[38] Whereas paint manufacturers patent specific products, and coding systems vary as discussed in Chapter 11, every base pigment has a consistent international pigment code. Pigment manufacture is a large-scale international business, owned by a few multinational companies. For example, phthalocyanide blue, which is a high-volume product, is made predominately in China and India, due to low labour costs. Other pigments, which are more complex to manufacture, are made in the UK, Europe and the USA.

Pigments and health

HEALTH scares were partly to blame for the collapse in the value of artificially manufactured dyes in the late nineteenth century. Arsenic, used to produce a particularly popular shade of green, known as Scheele's Green, was used in wallpaper in the 1860s, and was said to cause skin and respiratory complaints, to the extent that *The Times* newspaper noted that children sleeping in rooms decorated with the paper had died before the true nature of the malady was found.[39] Notoriously, it is thought that the wallpaper in Napoleon's villa on Saint Helena, where he was held in exile, may have contributed to his demise. Similarly, poisonous water was traced to dye stuffs released into watercourses from factories in Germany and the UK producing the new aniline dyes, and Victorians suffered skin irritations thought to originate from cheap versions of these dyes. The clamour to return to 'natural' dyes in the late nineteenth century was thwarted by the fact that the manufacture and the supply of dyes derived from plant and mineral extracts had largely collapsed, as the new artificial colours flooded the market. European legislation on the pigments industry continues to monitor pigments and solvents through EU Directives.[40] Reduction in the use of volatile organic compounds (VOCs) has generated industry-wide changes in manufacture and application. As the percentage of solvent has been reduced, the cost of paint has risen, but water-based paints are not yet thought to be sufficiently competitive in terms of performance to be able to replace substantially solvent-based coatings.

The general movement for 'natural' products, 'organic' food, and ecologically sound production methods has spurred a resurgence in pigments sourced from plants and minerals.[41] Technical data sheets for these products are readily available and, reassuringly, read more like a culinary recipe than an industrial product.[42] Manufacturers of these products argue that conventional paint will continue to release fumes into the atmosphere, which can cause respiratory problems. This plays on the irrational fear that associates chemical names with industrial processes, ignoring the fact that all natural materials are themselves chemical compounds. The range of colours in some ecological paints is, however, limited in comparison with the huge choice now common to consumers, and there is doubt about the validity of some supposedly 'eco' credentials.

Recently, there has been a growth in the use of traditional limewash, as used on the White House in Washington DC (1800).[43] At Stirling Castle, Scotland, the Great Hall (dating from 1503) was renovated in 1999 and rendered with a lime mortar and pigmented limewash.[44] Historic Scotland undertook a full archaeological study, and the evidence confirmed that a layer of lime render, which was pigmented either with red local sand or imported ochre, would have covered all of the stone rubble sections, together with a limewash painted over the ashlar stonework. The limewash is a natural consolidant, and forms a protective coating to the decorative ashlar. Visually, it unifies the different stone sections into a single architectural form. It proved to be a highly controversial move, however, as the new colour was very visible from a long distance and was a shocking change in appearance from what had come to be accepted as the 'authentic' grey stone.[45]

Pigmented materials and the future

ARCHITECTS have a responsibility to their clients to specify surface treatments and materials suitable to their location and use, and with appropriate maintenance requirements. It is therefore insufficient to choose products by colour alone.[46] Simple coloured renders, which make use of through-colour pigments, give permanence and, because they have thickness, can be considered as conceptually different to a thin surface coating, as seen on O'Donnell + Tuomey's Sean O'Casey Community Centre (2009) in Dublin. Pigmented materials – particularly cladding materials – such as fibre-cement, glazed terracotta, tinted, laminated, or back-painted glass, are increasingly used in contemporary architecture. The intensity of colour and durability make them attractive, as does their robustness as relatively maintenance-free surfaces.[47] Architects Allford Hall Monaghan Morris (AHMM) and Sauerbruch Hutton make use of glazed terracotta similar to traditional faience,

Pigmented render expressed as a thin skin on the Sean O'Casey Community Centre, Dublin, O'Donnell + Tuomey Architects (2008)

Square section ceramic rods as part of a layered facade at the
Brandhorst Museum, Munich, Sauerbruch Hutton (2008)

which was commonly used on facades, for example, by Hendrik Berlage at Holland House in London (1916). Contemporary usage tends to be as rainscreen cladding, which is highly durable and can take advantage of the vivid hues on offer.

Coloured glass, now available in a large range of colours, provides intense, clear colour by sandwiching a film of tinted plastic between sheets of glass. Alternatively, paint can be applied to the rear surface of the glass and fired. One problem with this product is the difficulty of predicting the effect of the self-colour of the glass on the resultant appearance and substantial investment in test panels may be needed if the end result is to be carefully controlled. Further developments in films with dichroic properties applied to glass, as used by Niall McLaughlin Architects in housing at Silvertown, London, and by UN Studio in the Netherlands, suggest a future in which colour is even more transient and unstable in appearance.[48] Nanotechnology is already producing new materials with unusual properties that may find a way into conventional practice. The artist Franziska Schenk is working with materials that are not pigments, but complex physical structures, which can reflect light, giving an iridescent, fluctuating colour.[49] They demand a new approach to art practice, as the colour is not immediately visible in the dull grey powder, but only appears once applied to a surface.

Colour, form and material surface

In architecture, colour choice is rarely dissociated from context or the material surface. It may be either applied or integral. Wall, floor and ceiling surfaces wrap and join to enclose space, or to define elements of form and suggest solidity. Recently, we have seen a resurgence of interest in surface, in decoration and applied finishes which allow for a more playful surface treatment. This dissociation of surface from form allows colour to be used in a more abstract manner, deliberately emphasizing the thinness of the coating. It is common for architects to have a general conceptual idea that a project will make use of colour, but then to isolate and select the actual hues at a relatively late stage in the design process. William Braham argues that colour is always subservient to form and notes that 'architects still close their eyes and wave their hands when the discussion turns to colour'.[50] Navigating one's way through this array of colour choice is a difficult task for lay people and professionals alike.[51] The relationship between form, colour and material surface is a complex one and the following chapters illustrate various approaches to the use of colour and how architects select the colours they use in their practices.

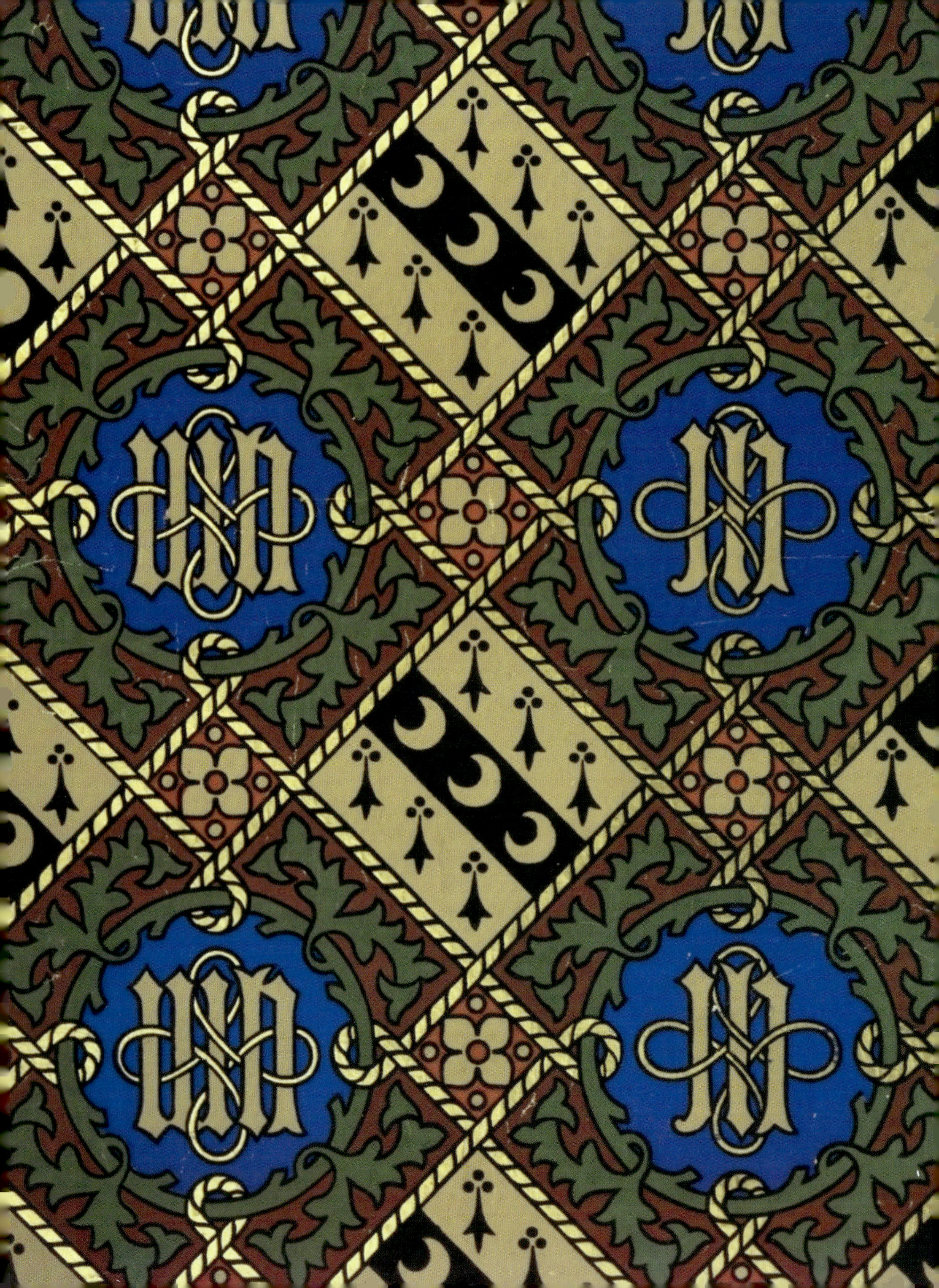

– 3 –

The unattainable myth of novelty: Caruso St John

Colour and cultural tradition

'The new, why always the new?' [1]

LONDON-based architects Adam Caruso and Peter St John are sceptical of the idea that architecture can or should be a product of invention, devoid of cultural and historical influence. Their work is highly eclectic, but, as they note 'in the good nineteenth century manner'.[2] Their projects, such as Nottingham Contemporary (2009) and the Victoria and Albert (V&A) Museum of Childhood in Bethnal Green, London (2007), along with their work with the artist Thomas Demand, are evidence of a recent fascination with colour and pattern.[3]

The range of pigments available to successive societies is contingent on technological innovation, yet the science of colour cannot fully explain the social and cultural traditions of its use. In the eighteenth century, it was common to pay homage to the rooms of antiquity by mimicking Greek and Roman motifs and colours, and, in doing so, to suggest that certain principles of decoration are irrefutable by virtue of such ancient precedent. Another common theme was the derivation of colour from a study of the natural landscape. This is evidenced in the work of Sir John Soane, for example, resulting in pale aerial ceilings, mid-coloured walls and deeper hues in the carpets.[4] Such iconic themes derived from nature appear throughout the eighteenth and early nineteenth centuries, recurring in the Arts and Crafts period in the late nineteenth to early twentieth centuries.

Caruso St John make studious use of ambiguity, complexity, pluralism and the picturesque. The sources for their architectural work are heterogeneous, making them difficult to characterize. They draw most consciously on cultural traditions – either from the everyday environs of a site, or from diverse historical sources. Indeed, Caruso St John seem to consciously search out ways to be unfashionable. Their office is understated, quietly tucked away in a nondescript area of east London. They dress unflamboyantly and have a strong interest in English Victorian architects such as Owen Jones, Norman Shaw, William Butterfield and Augustus Welby Pugin. Their work responds to the

facing page
Wallpaper design by Augustus Welby Pugin (1812–52)

writings of Gottfried Semper and Adolf Loos, particularly in respect of the relationship of cladding to structure. Observing the capacity of Islamic architecture's highly decorative and complex patterned surfaces to counterbalance compositions of dense volume, they have also started to explore the idea of decoration as being integral to surface. This interest in ornament as well as in colour is a relatively recent development in their work. The polychromatic patterns at the Bethnal Green V&A Museum of Childhood (2007) can be directly related to the work of Owen Jones, whereas the inlaid surface relief at Nottingham Contemporary (2009) makes reference to the work of the Chicago school of Sullivan and Adler, whose decorated frame buildings were themselves highly influenced by the work of Gottfried Semper and Karl Friedrich Schinkel.[5] Despite this pluralistic sourcing, their architecture is not remotely pastiche. Hal Inberg has observed that out of this attitude of sampling and re-mixing

> a surprising quality emerges. It is not heroic originality. Rather, it is a quality born of cultural literacy, a subtle yet restless formal intelligence and the political will to resist the cynical, laissez-faire contemporary forces that breed placelessness. In rejecting the modernist myth of originality, Caruso St John paradoxically free themselves to achieve that which they reject.[6]

Caruso St John refer to the 'unattainable myth of novelty'[7] as part of an ongoing critique of contemporary architecture that is embedded in their own writings. One result of the current ability to model – and subsequently manufacture – complex geometries not attainable prior to the advent of computer technology, they argue, is that architecture has again become obsessed with formal invention and originality. This criticism is aimed most directly at the production of extreme forms, some of which bear no relation to context physically, socially or culturally. In contrast, their preference is for an architecture that, while of its own time and social context, is not afraid to make reference to a continuum of thought and cultural tradition. They see no reason to reject the past, indeed their argument is that it is more radical to consider that which is known, and to place oneself within an ongoing and progressive cultural discourse. 'Now, more than ever, it is its cultural history that lends architecture continued relevance. It is architecture's capacity to be reflexive and critical that sets it apart from advertising, on the one hand, and pure science on the other.'[8]

What does this mean in relation to their use of colour?

Victoriana at Bethnal Green

THE colours most frequently used in the architecture of Caruso St John are pale greens, brick reds, muddy browns, deep aubergines and soft yellows. Colours are drawn from the surroundings of the sites, from the cultural and social history of the particular place and, frequently it would seem, from nineteenth-century England, a rich era in the history of colour, which also produced the most substantial leaps forward in colour theory, research, experimentation, and in the manufacture of pigments and dyes.

Michel Eugène Chevreul c.1860 (1786–1889)

Prior to this period, pigments were predominately natural ground minerals or extracts from plants, insects and other organisms.⁹ The sudden advances in chemistry in the mid-nineteenth century resulted in pigments that were less susceptible to deterioration in sunlight. Many of the major texts on colour theory were written by chemists, perhaps most notable in this context, by Eugène Michel Chevreul, who retired from his position as head chemist at the Gobelins tapestry company in France at the age of fifty-three, and then spent the next half century of his very long life investigating colour. His book *The Laws of Contrast of Colour: and Their Application to the Arts of Painting, Decoration of Buildings, Mosaic Work, Tapestry and Carpet Weaving, Calico Printing, Dress, Paper Staining, Printing, Illumination, Landscape and Flower Gardening, etc.* (1839) was highly influential and was intended to give practical advice across this wide range of applications.¹⁰ There is no doubt that Owen Jones, the English architect, best known for his contribution to Joseph Paxton's Great Exhibition building of 1851, was aware of the book as he makes reference to it in his own seminal book *The*

Grammar of Ornament (1856). Jones had a mission to educate his contemporaries in the principles of colour theory but also, by cataloguing historical pattern and architectural decoration, to prevent an unenlightened plundering of ornament merely as fashion.[11] Owen's book begins with a series of 'Propositions', the second of which is that 'architecture is the material expression of the wants, the faculties, and the sentiments, of the age in which it is created'. This chimes with the attitudes of Adam Caruso and Peter St John. Although taking a highly scholarly approach to historical sources, they insist that their architecture is inevitably of its own time. It does not merely mimic; it samples, mixes and re-appropriates colour, decoration and intent from a very wide range of antecedents.

During the nineteenth century, scholars began to realise that classical Greek buildings had originally been polychromatic. As Victoria Finlay notes:

> Doric columns were striped red and blue; the Ionic capitals sported gold as well. The findings were not generally welcomed. There is a story of the sculptor Auguste Rodin once famously beating his chest: 'I *feel* here that they were never coloured,' he proclaimed passionately. Colour historian Faber Birren recounted an anecdote in which two archeologists went into a Greek temple. One climbed to the cornice and the other yelled, 'Do you find any traces of colour?' When he heard an affirmative answer he howled 'Come down instantly!'[12]

Extension to the V&A Museum of Childhood, London,
Caruso St John Architects (2007)

Detail of stone facade, V&A Museum of Childhood, London,
Caruso St John Architects (2007)

Owen Jones' colour scheme for the Great Exhibition building (1851), known as the Crystal Palace after its move to Sydenham, was therefore highly controversial. He applied his own colour theory in using primary colours of red, yellow and blue high up on the structure, with secondary colours lower down. In Caruso St John's V&A Museum of Childhood, in Bethnal Green, London, designed in collaboration with artist Simon Moretti, a primary red is used in intertwined circular patterns on the otherwise white ceiling of the main exhibition room. It makes reference to the highly decorative patterns on the original mosaic floor of what was the Bethnal Green Museum (1872), which was laid by women prisoners. Elsewhere, in the new entrance hall, brown, green and gold (a shade of yellow) are used to contrast with the polychromatic stone facade, which uses a palette of red quartzite and brown porphyries. The colours are reminiscent of those favoured by the Gothic revivalist Augustus Pugin, who used a palette of gold, deep reds, olive green, pale yellow and, occasionally, ultramarine in his wallpaper designs.

The new dyes and pigments offered greater stability and variety than had previously been possible. In textiles, colour trends can be fickle and subject to wide and sudden changes in fashion or manufacturing ability. William Perkin was a young student charged with trying to isolate synthetic quinine from coal tar to treat malaria. He stumbled on an intense purple colour in the residue from an unsuccessful experiment in 1856. Unlike many who would have thrown it away, Perkin seems to have immediately understood its potential. The unpromising solid dye was developed with advice from Robert Pullar, a Scottish dye manufacturer (and subsequently dry cleaner), based in Perth. Perkin named the colour 'mauve' and made his fortune from sales of the dye, retiring to the country at the age of 36.[13] Perkin's breakthrough was hugely significant in the history of colourants and, subsequently, to a vast number of chemical advances in plastics and, ironically, medicines. In his book *Mauve*, Simon Garfield parallels the huge growth in aniline dyes with the gradual acceptance by the university communities of the study of chemistry as a subject in its own right. In Perkin's case, the suggestion that an academic would sully himself with an industrial process that would make him rich, and would embrace the lowbrow social and cultural obsession with purple, was frowned upon by his more senior colleagues. Fashion, seen almost exclusively as a feminine pursuit, was (and still is) continually associated with frivolity and passing fads. As the satirical magazine *Punch* noted in 1859:

> London is in the grip of mauve measles, spreading to so serious an extent that it is high time to consider by what means it may be checked … and from the effect which it produces on the mind contend it must be treated as a form of mild insanity. … Although the mind is certainly affected by the malady, it is chiefly on the body that its effects are noticeable.[14]

It goes on to describe the disease as infectious, starting with 'a measly rash of ribbons', noting that 'mostly women are affected but men could be treated with one good dose of ridicule'.[15]

facing page
Interior of entrance foyer, extension to V&A Museum of Childhood, London, Caruso St John Architects (2007)

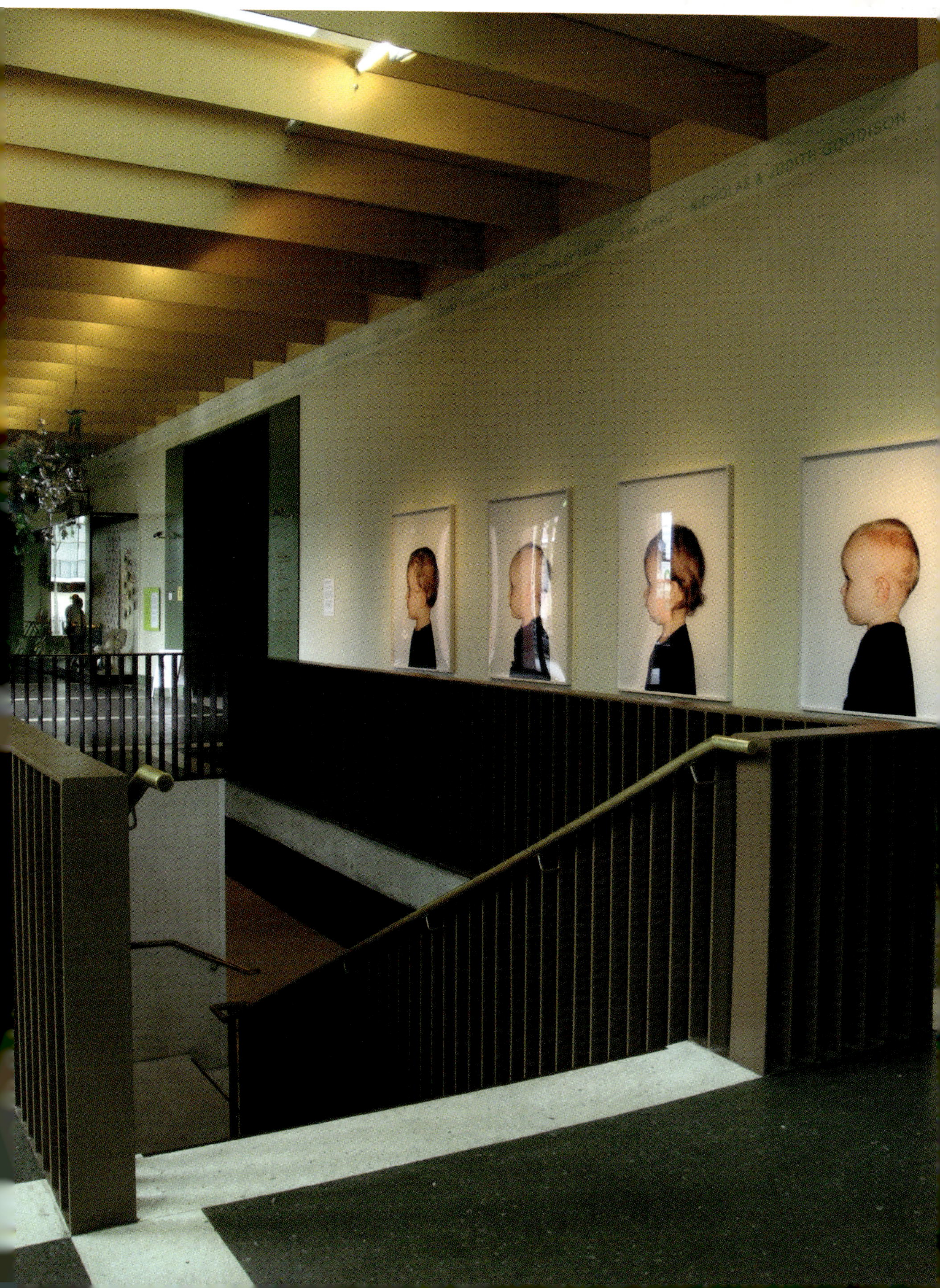
NICHOLAS & JUDITH GOODISON

Fashion and anti-fashion

SUCH short-lived fashions, while easily accommodated within the transient clothing industry, present architecture with more of a dilemma.

The key protagonists of early twentieth-century Modernism sought to wipe the architectural slate clean and to develop an architecture with minimal reference to precedent or historical association. Originality was the goal and to admit to using cultural history as a prime source was considered a crime. As Modernism kicked at the heels of Art Nouveau, deriding the style as fashionable, feminine and essentially short-lived, the new age projected itself as anti-fashion and timeless. Adolf Loos made a comparison between facade cladding and dressing and extended this to his own attitude to clothes and to fashion 'to be dressed correctly … is to be dressed in such a way that one stands out the least'.[16] The 'English style' of the time was considered by Loos as exemplary, being essentially modern, sophisticated and understated. Despite the Modernist obsession with newness, with stripping bare, with clean lines and smooth and undecorated surfaces, one is aware that to apply a white layer is a positive choice and is itself a coating – albeit a very thin one, masking the construction. Unlike later 'High Tech' architecture, there was little desire in Modernist architecture to expose the structure – or the skeleton, or to wear 'clothing which represents the structure that it covers … but … minimal clothing, a decent cover … a discrete mask'.[17] Partly because black and white images were the predominant medium through which the architecture was communicated, the architecture of the early twentieth-century Modernists has been frequently associated, often wrongly, with monochrome, white and grey buildings. Not only are the images deceptive, but commentaries from the period suggest a pervading colour blindness, a refusal to acknowledge the presence of colour and confused and often contradictory positions taken by some of the key Modernist architects themselves. Mark Wigley's book *White Walls, Designer Dresses* exposes a number of such paradoxes in the writings of Loos and those of a contemporary, Herman Muthesius. Wigley notes an implicit understanding of the 'double function of fashion – as mask and marker', hence the surface dressing bonds individuals to a group and detaches one group from another. In fashion, haute couture aims to express distinction, yet quickly becomes diluted and assimilated as orthodox high-street copies. Fashion must constantly steer away from the norm or face becoming blandly ubiquitous. Ironically, social groups that yearn to express unconformity can, paradoxically, wear uniform clothes in support of their shared identity, such as the goths, mods or Hell's Angels. Just as Adolf Loos argued that a gentleman should dress in a manner that does not stand out, there is an implicit understanding that such restraint in dress is itself a statement of social standing and distinction. The metaphor of clothing, which is intertwined with fashion, suggests that one could trace both an underlying interdependency between colour in architecture and the timeframe in which it is conceived, and also a contrary ambition for anti-fashion, namely a sense of timelessness.

The British clothes designer Jean Muir preferred to use the word 'fashion' as a verb and was highly successful in negotiating the tightrope between fashion and timelessness. Her understated clothes were highly crafted, with the form, cut and the cloth being paramount. She is remembered most for navy blue, timeless classic couture, although she would also work directly with the dyers and textile manufacturers to seek particular seasonal colours. The historical availability of particular pigments and dyes through technological advances has had an influence on the predominance of certain hues, and the emergence of new ones, in both textiles and pigments. But, beyond technology, there is less certainty as to why specific colours are associated with particular eras in relation to the use of colour in interiors, in textiles and in architecture as a whole. A whole industry exists based on colour prediction – the basis of much interior design, fashion, textiles and household paint is more or less dependent on the colour forecasters who periodically predict new shades and fashion magazines that define what will become the 'new black' for the season. Yet it seems that there is little in the way of empirical research on which to base these predictions. It seems to be 'a matter of rationalizing decisions based on part experience, observation and intuition'.[18]

Fashion, being essentially cyclical, can frustrate colour theory, as what is perceived to be acceptable fades then returns.[19] Indeed, cynics would suggest that the entire industry is set to perpetuate consumerism by encouraging the homeowner or the fashion-sensitive to believe that they really must redecorate or buy the colour of the season. Fashions feed on those who are swift to adopt a product or to pick up an emerging trend in colour. According to Malcolm Gladwell's theory of the 'tipping point', the sequence that will generate a consumer epidemic starts with the innovators who take risks, feeds on the 'early adopters' who set the trend and becomes established by a surge of followers until a majority is reached. Paradoxically, only those who refuse to follow suit or who are very late in catching on then stand out from the crowd. Is this also true for trends in colour?

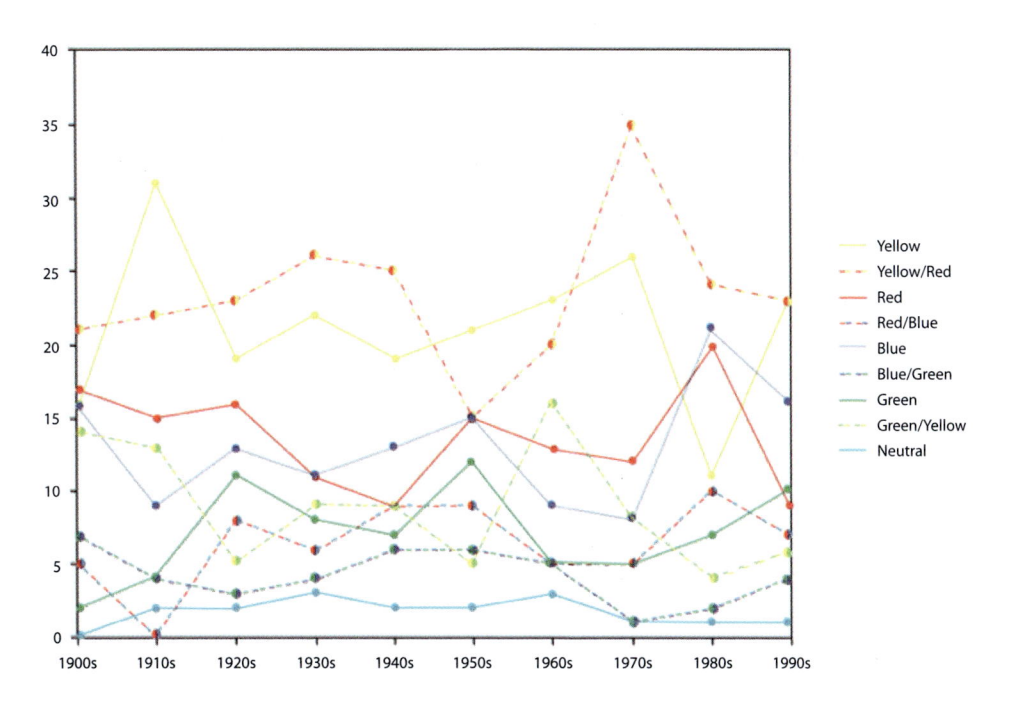

Graph of Natural Colour System (NCS) hue categories as a percentage of each decade's colour palette

'Does a colour become the hot thing because a colour expert says so, or is it because enough people told him about it first?'[20]

Opposing attitudes prevail in respect of colour chronology. Those who point to commonly held principles of colour psychology will consider that these timeless principles must overrule any trend. 'Short-lived variations in interior design follow a hasty, disposable mentality, and contradict serious and fundamental design philosophies.'[21] Proponents of this philosophy tend to consider the whole issue of trends to be an ineffective way of designing using colour. Others, however, will readily absorb colour trends, whether consciously or subconsciously. A few studies have tracked the use of particular colour through the decades with certain variations clearly evident, but none that would suggest that prediction can be based on previous use in any scientific manner. Nevertheless, colour forecasting has established itself to the degree that an exhibition *Archeology of the Future: 20 Years of Trend Forecasting with Li Edelkoort* celebrated the work of one of its best-known advocates.

Reflecting and documenting past use is somewhat easier and, by using magazines, trade journals and colour cards, Australian researchers Stansfield and Whitfield have plotted trends within their continent by the decade throughout the twentieth century. Their research also plots the Natural Colour System (NCS) levels of relative 'blackness', 'whiteness' and 'chromaticness' predominant in each decade. Chromaticness peaks in the 1970s, at the same time as whiteness drops. This would indicate a general level of colourfulness and minimal pastel shades at this time, in Australia at least. The most consistent performers across the twentieth century are the yellow-reds, with yellows peaking in the 1910s and oranges in the 1970s. The 1980s sees the highest peak of whiteness, suggesting a predominance of pastel colours along with the highest use of red-blue.

One may, for example, associate lime green with the 1960s, orange with the 1970s, and a shift to browns, creams and neutrals by the end of the decade and into the 1980s. Tom Porter, who has written extensively on colour in the UK, observed similar fluctuations in colour use. A number of UK-based contemporary paint companies offer 'Heritage' ranges which are, they suggest, borne out of research on period colours such as paint scrapes taken from historic buildings, be they Victorian deep reds, greens and gold or softer Georgian greens and pale blues.

Although individuals will always have preferences on colour, there is clearly an underlying zeitgeist borne out of popular culture, which can have the effect of moderating the palette of a society at a given time. Rem Koolhaas has noted that 'The future of colours is looking bright'.[22] The use of colour and the discussion of colour in architecture has, without doubt, been increasing in recent years. Fashions in relation to colour, and colour trends, are undeniably influential. While some people will strive for difference, some seek to conform, wishing to be seen as being in tune with fashion. Others are content to develop an identity independent of passing fads, or specifically to seek endurance.

facing page
The New Art Gallery, Walsall, England,
Caruso St John Architects (2000)

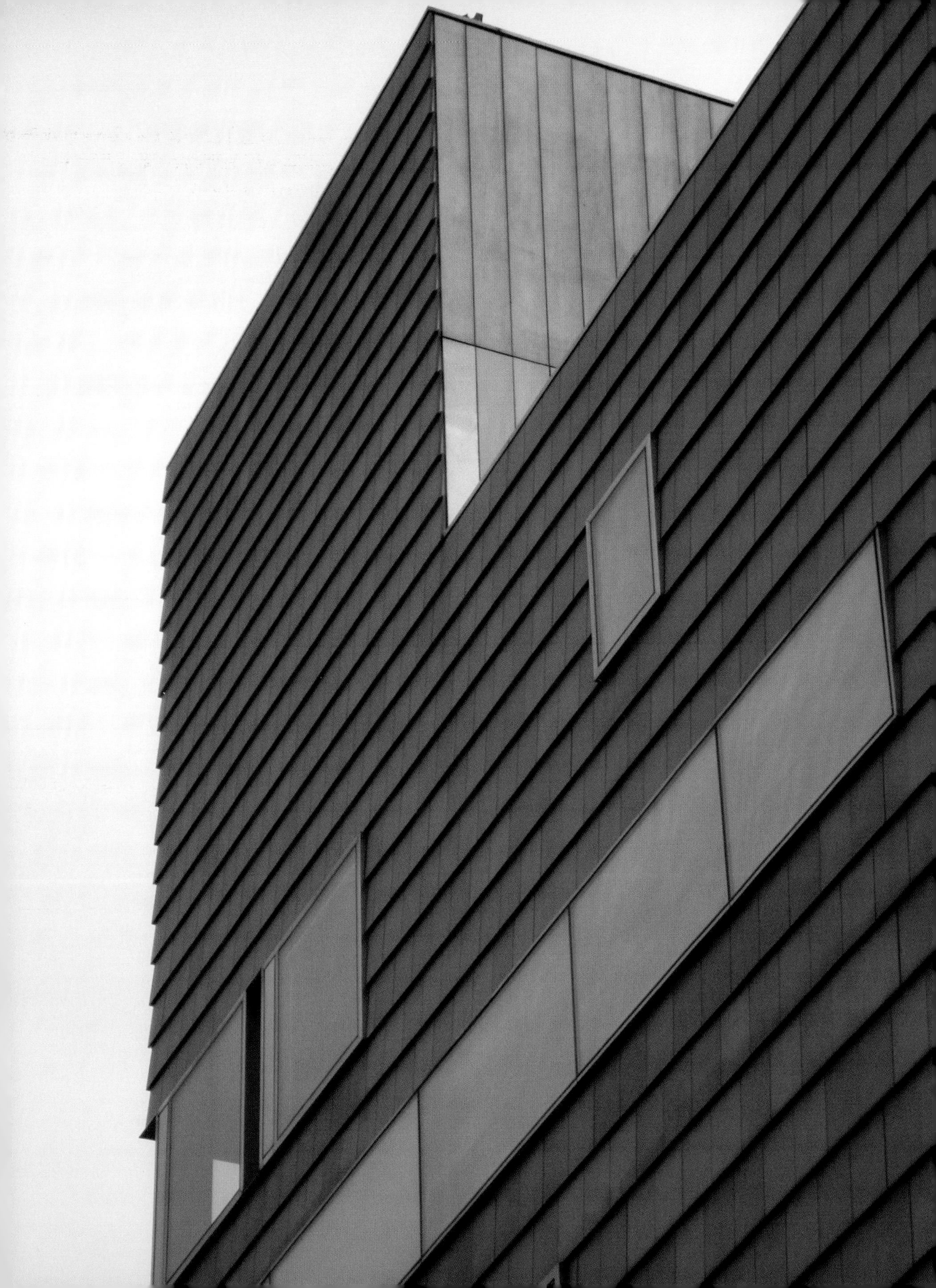

Dressing and wrapping

WITHIN their pluralist, heterogeneous approach there are some constant principles in Caruso St John's architecture. Threading through their work is an ideology that considers the atmosphere of a place or room as paramount, and that all other devices – the means of production, materiality, structure, construction, pattern and colour – are subservient to the feeling and character of the space. Rooms are predominately made by enclosing walls, rather than by a frame, which they consider gives a greater capacity for ambiguity than is afforded by columns. The actual surface of the wall – its materiality, texture and colour – is manipulated to invoke an emotional response.

The part played by colour and pattern in their architecture has gradually evolved as part of this language. Adam Caruso and Peter St John trained at McGill University in Montreal and at the Architectural Association (AA) School in London, respectively. Caruso St John's early work was centred on the ordinary, on found contexts and found materials left in their natural state, such as taped plasterboard, floated concrete screeds and fair-faced brick. Their interest in colour can be traced to this earlier interest in tectonics and the communicative potential of materials. They have a genuine interest in contemporary art and their knowledge and sensitivity were among the reasons cited by the commissioning client for their first major project, the Walsall Art Gallery, completed in 2000.[23] Their focus is on providing gallery spaces that do not compete with the art, but have a very specific character. The Walsall building appears stark and modern in its setting, like an inflated Adolf Loos house, and similarly awkward in its composition and elevational treatment.

The comparison to Loos can also be usefully applied to the relationship of the internal spaces as a series of rooms or volumes intertwined by a complex circulation that wraps around the edges of rectilinear spaces. This provides opportunities to view the art from different perspectives and in changing juxtapositions. In describing the gallery, the architects themselves use the analogy of a large, rambling house.[24] The board-marked concrete walls are lined in key places with a second skin of Douglas fir, defining rooms within rooms and wrapping the spaces to define the individual atmosphere of each space. Construction is subordinate to the surfaces of the container, and is manipulated as part of an aesthetic control of space. This does not pretend, however, to be an 'honest architecture'.[25] There is no truck with any dogma that would demand that the structure must be exposed and must relate directly, or be essentially true, to the space and plan. The surface is akin to a mask, or to a layer of clothing, prudently covering the body, in some cases accentuating or correcting structural necessity for the sake of aesthetic control.[26]

The Austrian architect Heinrich Kulka coined the phrase 'Raumplan' in 1931 to define the spatial order evident in the work of Adolf Loos, where the plan is only part of a three-dimensional composition.[27] Spaces are interlinked in section and although they appear contained by walls, the contiguous volumes merge. Loos' 'Raumplan' houses, such as the Müller House in Prague (1930),

facing page
Interior, the New Art Gallery, Walsall, England,
Caruso St John Architects (2000)

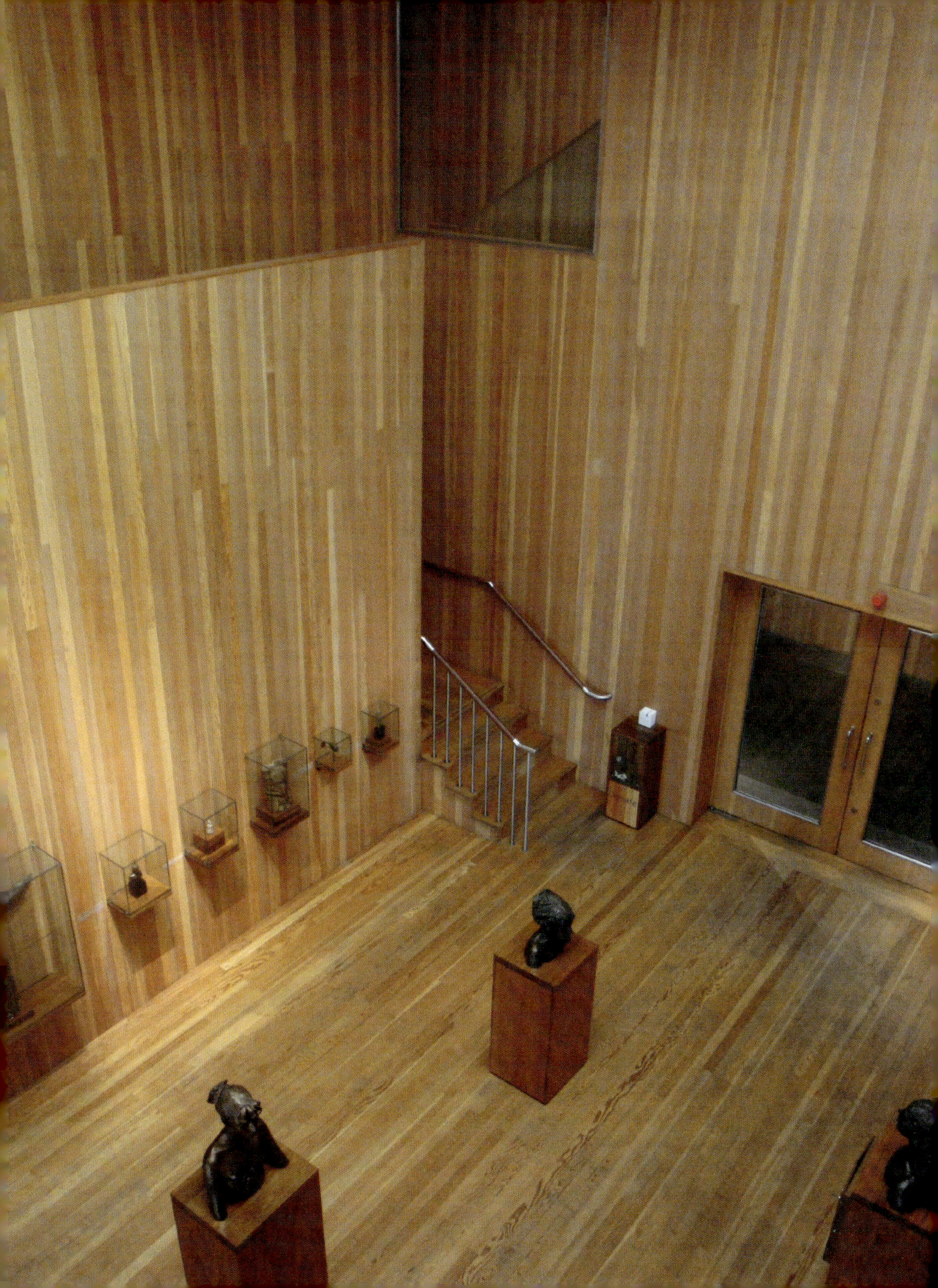

further define rooms within rooms by the use of colour and other surface cladding treatments, such as the green walls and varnished veneered wood of the breakfast room.[28] Elaborating their interest in Loos, Caruso St John explain that:

> Loos's essay, ... 'The Principle of Cladding' is one that we have discussed with students for many years, especially from the position of being architects in Britain, where architecture has been for so long obsessed with the construction of the wall and forgetting about the character of the room. Loos's prescription of what not to do is a perfect description of British high tech architecture. We definitely subscribe to his position, but are interested in the matter of the walls, in how we can make their construction somehow responsive to a spatial imperative.[29]

Although colour plays no significant part at Walsall, the principle of the spatial organization and layered meanings are constant themes.

In 2009, Caruso St John subdivided Mies van der Rohe's Neue National Galerie in Berlin with vast, thick woollen curtains. Instead of being seen as open plan, or a traditional white-walled sequence of spaces, the curtains introduce ambiguity, defining rooms within the heavy steel frame and acting as backdrops for an exhibition of the work of the artist Thomas Demand. The fabric uses soft colours, slightly muddy yellow, grey green and browns to dress the space.

The use of cloth hangings to define enclosure within a frame can be interpreted as a reference to the German architect Gottfried Semper. In his taxonomy of architecture, Semper cited the origin

Thomas Demand Exhibition, Neue National Galerie, Berlin, Caruso St John Architects (2009)

of wall-dressing 'Wandbekleidung' as carpets hung over a frame around the hearth of primitive dwellings to define space.[30] Semper argued that 'the wall should never lose its original meaning as a spatial enclosure by what is represented on it'. Decoration, pattern and, in this case, colour can therefore be applied to the wall surface, but the fundamental role of the wall is as enclosure. Caruso St John go further to argue that the first priority of the architect is to set the atmosphere or character of the space in order to evoke an emotional response. In other words, although in the case of Berlin the frame was provided by Mies van der Rohe, one should start by imagining the curtains, then devise a structure to support them.[31] The use of colour in this process first emerged in a refurbishment project for national offices of the Arts Council England in London (2008), where they worked in collaboration with the artist Lothar Götz. In the resulting design, colour is used to wrap space, predominately in the vertical planes of walls. In some cases, such as the boardroom, multi-coloured panels disrupt the integrity of the walls, giving prominence to the surface over the form but give a very distinctive identity to the space.[32] This polychromatic boardroom is the culmination of a sequence of key spaces that suggest a hierarchy within the refurbished offices, utilizing seven strong colours to provide identity and character against the general background of a warm stabilizing grey tone. The intense colours, perhaps influenced more by Götz, have given way to a more subtle palette in later projects.

Such concepts of wrapping and dressing are exemplified in their most significant project to date. Completed in 2009, Nottingham Contemporary is a large art gallery set into a slope in the centre of Victorian Nottingham. The industrial heritage of the area was lace-making, and the built context is one of red brick and pale sandstone. Rather than employing the same materials, the building sets up a contrast using a green-tinted concrete, black-tinted concrete and gold paint. The building is not immediately likeable; it is intriguing, puzzling and unsettling. Perhaps we are simply not used to buildings being green, at least not green on the outside. Adam Caruso admits, however, that they had been trying to design a green building for some time. Their competition entry for the Zurich Landesmuseum sported a tiled frieze and green textured concrete panels,[33] as does another unbuilt project for the Politiegebouw in Kortrijk in Belgium (2007).

The application of pattern, in the case of Nottingham, derived from lace laid into the moulding process, is also not commonly associated with a contemporary building. The architects are making cultural references, not only to the direct relationship with the nearby Lace Market, but also to the work of Dankmar Adler and Louis Sullivan of the Chicago school.[34]

Green-tinted, lace-printed concrete panels against the Victorian red-brick surroundings of Nottingham Contemporary, England, Caruso St John Architects (2009)

Scalloped panels (top) and green
and gold (right) at the Nottingham
Contemporary, England,
Caruso St John Architects (2009)

As noted earlier, Caruso St John deride the undisciplined application of technology in search of originality of form, such as using three-dimensional pattern-making by employing shifting fractal geometries. Their forms, like those of Adler and Sullivan, tend to be simple. Their use of cultural and historical precedent, and of colour and pattern, evolves constantly and is employed to maintain their idiosyncrasy. In support of the application of two-dimensional pattern, they refer to the huge investment that societies have made in developing geometrical and other ornamental patterns with which to embellish buildings. Indeed, it is a twentieth-century phenomenon that architects ceased to feel comfortable with surface decoration. Surface patterns and textures are now beginning to re-emerge, however, as technologies are developed to replicate, emboss and etch images without recourse to the craftsmanship that now seems lost, or is prohibitively expensive.[35] The latest nineteenth-century industrial technology was also employed in the structure and construction of Sullivan and Adler's architecture. The facades were then decorated with ornamentation symbolic of organic forms.

At Nottingham Contemporary, the making of the concrete scalloped facades, reminiscent of fluted Ionic columns but which are non-structural, employed a complex modeling process, using digitized computer numerical control (CNC) patterns cast first into latex, and then into concrete, giving a creamy and embossed texture. One theme in their work – the desire to make ordinary materials extraordinary – is exemplified in the building. The pattern is one thing, but the combination of pattern and green colour, with the vertical stripes of gold set between each panel, is the most memorable aspect of this building. In common with fashion, Caruso St John will happily re-interpret, mash and mix, but the result is of its own time. As Laura McLean-Ferris observes of the building, 'A romantic lacy Victoriana, the demure ruffles and folds of an ankle-concealing skirt, and the industrial factory spaces in Nottingham that created such garments, are recalled'.[36]

The use of gold on the building evokes further association. Located adjacent to a church, the gold on the stripes and the upper sections of the facades can be seen as provocative as they gleam intensely in the sunlight, apparently competing for attention. Although Pugin used gold frequently in his wallpaper designs, the fragmented form and geometry of the building are reminiscent of Hans Scharoun's Berlin Library and Chamber Music Hall. As with the use of green, a gold facade is unexpected. The light conditions modify the upper gold sections from muted to brilliant over the space of seconds.

Caruso St John consider themselves equally capable of being 'Gothic' and 'Classical'. Nottingham Contemporary combines

Nottingham Contemporary in its wider context, Caruso St John Architects (2009)

Interior detail, gloss-painted toilets at Nottingham Contemporary, England, Caruso St John Architects (2009)

the Modernist board-marked concrete, white walls and pale floors reminiscent of the Walsall Gallery, with an exuberant, lacy surface and intense colour. Internally, the combination of gothic and classic is most refined. The Gothic, or perhaps in this case more a Romantic, impulse is evidenced in the use of soft, fluffy cranberry-coloured soffits in the offices and café, tempering the acoustics of these spaces, combined with turquoise and pink patterned curtain fabric evoking the work of 1950s' textile designers such as Roger Nicolson. Intense purples and pinks in highly reflective gloss paint are used in the toilets and a deep aubergine beneath the cantilevered entrance. These are set against the classic Modernist pure white walls of the galleries and exposed concrete and stainless steel. This odd, playful combination partly explains the difficulty of pigeon-holing the architects. Frank Werner suggests that the history of architecture has been 'an incessant struggle between two alternately successful factions, between closed and open spatial forms and formal concepts, between Dionysian and Apollonian formal approaches'.[37] The struggle between the Gothic and the Classical, Werner notes, is not purely a stylistic battle, but is essentially a struggle between opposing theoretical models. To lay claim to both gives Caruso St John a very rich provenance for their work and allows them the freedom to experiment with colour, texture and pattern in pursuit of an emotional response.

Caruso St John believe that 'architecture is fundamentally stable, making places whose character is informed by material, size and location'.[38] They are not drawn to the need for complete originality, but do seek distinctiveness. The use of colour in the architecture of Caruso St John is still developing. Despite their belief in the stability of architecture, recent projects suggest a subtle shift away from the found off-the-peg materials of their early projects to a bespoke, tailored and crafted 'clothing' that is

Café interior, Nottingham Contemporary, England, Caruso St John Architects (2009)

not afraid to use colour. Architecture demands a longevity that is simply not necessary in clothes and so the analogy with dressing can only be taken so far. While we can readily transform the atmosphere of spaces with a simple layer of paint – whitewashing, erasing, invigorating, calming, protecting – or with the use of textiles to enclose and wrap, colours that are embedded in pigmented materials, in ceramic tiles, tinted concrete or coloured glass panels must be designed and made to endure.

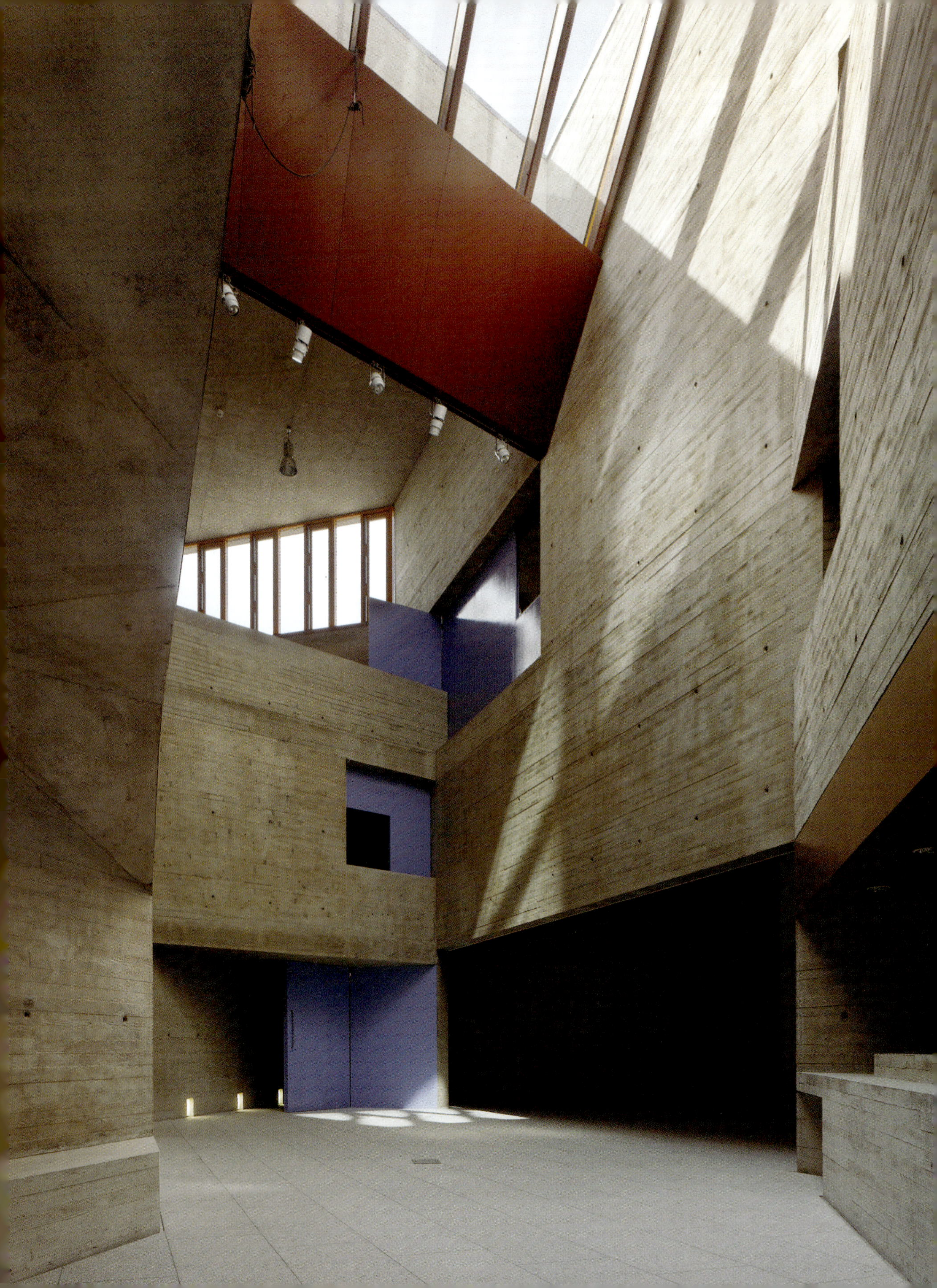

– 4 –
An intuitive palette: O'Donnell + Tuomey

O'DONNELL + TUOMEY are based in Dublin, Ireland. Although the practice, established in 1988, has fluctuated in size, there is consistency in the office's output – a gentle, thoughtful, well-considered and crafted architecture. Tod Williams and Billie Tsien make reference to a sense of sobriety, to restraint and discipline in the office's work.[1] The relationship between the practical and the poetic underpins their work, as Tuomey acknowledges, using a quotation from Seamus Heaney:

> I wanted to affirm that within our individual selves we can reconcile two orders of knowledge which we might call the practical and the poetic; to affirm also that each form of knowledge redresses the other and that the frontier between them is there for the crossing.[2]

Both Sheila O'Donnell and John Tuomey worked in the office of James Stirling on the Staatsgalerie in Stuttgart, and were based in London for a number of years, before returning to Ireland in 1981. The office of Stirling and Wilford consistently used colour in its work, often juxtaposing strong colours in controversial fashion, although executed in a very different manner to O'Donnell + Tuomey's work.

facing page
An Gaeláras, Cultural Centre for the Irish Language, Derry, Northern Ireland, O'Donnell + Tuomey Architects (2009)

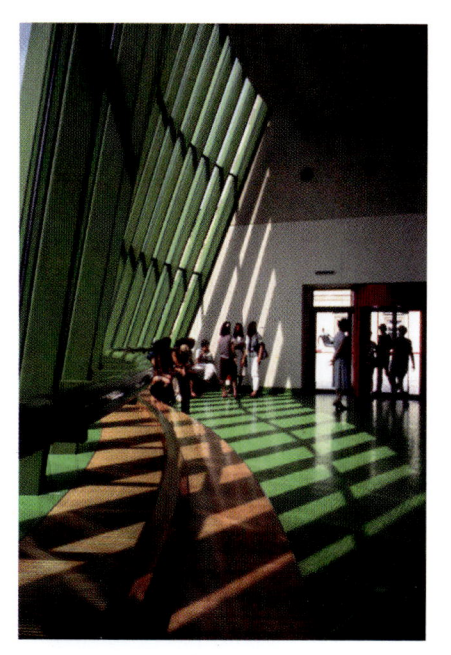

Neue Staatsgalerie, Stuttgart, Germany, James Stirling (1984) (completed by James Stirling, Michael Wilford and Associates)

Colour in the ethos of the office

Colour is not an obvious theme in the work of O'Donnell + Tuomey, yet it is present, rigorously considered and confidently expressed. A consistent aspect of their approach to design, according to O'Donnell, is to 'try to unlock something about the place, the site … the essential nature or character of it'.[3] Thinking strategically about a plan, or the section of the building on the site, they will simultaneously consider the atmosphere and character of the building. The pragmatic and experiential aspects are developed in parallel. Early on in a project, therefore, there may be a conceptual idea that colour should be used in the development of the character, but without a commitment to any specific hues. Unlike some of the other architects under consideration, their use of colour is not entirely project specific. There is consistency in the way in which colour is used conceptually, which suggests a set of principles underlying the work. Although these principles ground their architecture, and their use of colour, Williams and Tsien also suggest that 'there is a unique syncopation between thought and intuition, the interplay between these two strengths – propriety and impropriety'.[4] In their use of colour, a similar method emerges. There is an enthusiasm for a greater depth of understanding about colour, through reading and experimentation, yet this is tempered by intuitive and subjective decisions. They experimented with a colour system at Adamstown, Dublin, where masterplanners Metropolitan Workshop, based in London, had brought in Grete Smedal as a colour consultant on the project. Smedal used the Natural Colour System (NCS) rigorously at Longyearbyen, Norway, and O'Donnell + Tuomey, who were architects for part of the Adamstown project, became interested in exploring the way in which the NCS system could, in theory, provide harmonious families of shades by using codes to group or contrast.[5] O'Donnell notes that, although they tried this, they were not convinced by the outcome and made final adjustments in the exact shade of colours, judging by eye.

Tuomey makes reference to the role of such intuition in respect of the Glucksman Gallery in Cork. It was an unusual project for them in that the site was extremely sensitive, as part of a picturesque landscape in a highly prominent position. 'The expectation was explicit, the architecture was part of the brief. … We had to think differently and make an intuitive, comparatively unsupported first step.'[6] This leap of faith has produced perhaps their most significant building to date, which was shortlisted for the Stirling Prize in 2005. Tuomey is, however, quick to qualify intuition as being distinct from chance, quoting from Agatha Christie's detective Hercule Poirot:

> a guess can either be right or wrong. If it's right you call it intuition. If it's wrong you don't speak of it again. … What is often called intuition is really an impression based on logical deduction or experience.[7]

O'Donnell + Tuomey's approach to colour has matured as their work has progressed, through developing an understanding of its role, and through judgement based on experience. If a colour works well, why not use it again?

O'Donnell + Tuomey's range of colours, or palette, is comparatively restricted. O'Donnell acknowledged that key colours are taken from their observation and sketches of Italian paintings and frescos from the Renaissance and earlier. Seeing the combination of a limited number of strong colours in the work of Fra Angelico (1395–1455) was a formative experience, from which four opaque shades of earthy red, yellow ochre, paleish green and a purply light blue had an intuitive resonance with the architects and were adopted. These have been offset by deeper browns, dark greens and greys. The fact that the original colours were natural mineral pigments was important; they had a sense of being dug from the earth and resonated with O'Donnell + Tuomey's grounded approach to architecture. O'Donnell + Tuomey's work has become symbolic of a type of critical regionalism, as part of an expression of a generation of contemporary architects who retuned to Ireland to work and whose architecture is identified with, and defined by, their country.[8] That O'Donnell + Tuomey have adopted colours sourced from Italian culture, over which they acquire authorship through use, demonstrates a confidence and antiparochialism that is evidence of the 'subtle bursts of sensuous will' which give an emotional spark to their architecture.[9]

Fra Angelico, *Altarretabel von San Domenico in Fiesole, Szene: Thronende Madonna*, detail, 1424–25

Colour, form and surface

THE end wall of O'Donnell + Tuomey's office in Dublin is stacked with a shelved archive of architectural models. Their models are extremely consistent – always in grey card, with a real aversion to using colour in a model:

> models are strange beasts ... they are useful for examining form and space, solid and void, light and dark, but if you try to be literal about surfaces in models, ... they become like a doll's house ... and you can lose them.[10]

Their models are therefore dematerialized, deliberately homogenized studies. Materiality emerges through studies utilizing drawings, watercolour paintings and sketches. Although the watercolours introduce a vivid sense of colour, they generally take the form of abstract blocks of pigment. They are used to clarify the form and conceptual framework, or the essence of the site or natural landscape,

Johannes Itten, *Vorkurs* or
basic curriculum at the Bauhaus (1923)

facing page
Strongly coloured terraced houses (top)
surrounding the site of the Sean O'Casey
Community Centre, Dublin (bottom),
O'Donnell + Tuomey Architects (2009)

but are non-specific in terms of any level of detail. O'Donnell's watercolours are used in a similar way to those of the American architect Steven Holl. The colour begins to be representative of early ideas of materials, such as for their housing at Adamstown, Dublin (2008), where brick solids are depicted, with recesses carved into the form.

In their search for character in architecture, Tuomey considers the relationship between 'form, substance and appropriation by use' to be key components.[11] Responsibility for form is seen as being firmly in the hands of the architect. Architectural form, to them, is never abstract or solely sculptural, however, but directly related to materials, construction and use. Johannes Itten's *Vorkurs*, the basic curriculum developed for the Bauhaus, considered colour as an equal seventh of the curriculum and, critically, placed it as a material alongside stone, wood, and glass. O'Donnell, although not making any direct reference to the Bauhaus, adopts a similar position: 'a material like brick is almost like a solid pigment … it is a colour-infused material … we think of colour as almost being like a material'.[12]

At the Sean O'Casey Community Centre (2009), colour is applied to only one face of a towering concrete structure, which announces the centre from a distance. The clients were conscious of the rapidly developing docklands that surround the site and wanted to have a building with presence, to balance an existing church tower nearby. The small terraced houses that surround the site are very colourful and were a direct influence. The area is known as the East Wall as it faces the harbour and the sea. The colour on the east face developed from an idea that this facade would take as its context the colours of the sea. Early design paintings showed it as blue, and this developed into a greenish colour. The colour is used in an abstract manner as a clearly expressed layer, with the thickness of the render expressed as a plane at the corner. There is no attempt to suggest solidity in the colour, it is shown for what it is – a single surface on a solid object, as illustrated in Chapter 2.

By contrast, the social housing at Galbally, County Limerick (2000–02), uses colour to enhance the form by applying colour to recesses in the elevations. O'Donnell + Tuomey tend to favour durable, natural, self-coloured materials externally. Applied colour, therefore, is more associated with the interior in their conceptual approach. At Galbally, the recesses are punched into the solid form to reveal a sense of interiority on the exterior. Recesses are a common device in their work, such as in the Ranelagh School and in the brick cubic forms of Adamstown. By adding a colour to the recess, O'Donnell notes, 'you are adding an effect on these two planes and their relationship with one another. So it's a spatial thing, to an extent.' The effect of colour to make some elements appear to recede and some to advance is consciously applied by Le Corbusier at Pessac, and at the Priest's House at Ronchamp.

The coloured recesses at Galbally suggest a blurring of the threshold of inside to outside to make the transition into the low-budget dwelling richer and more complex. O'Donnell notes that they are intrigued by the broken patches of colour used by Le Corbusier in projects such as the Unité in Berlin (1959), in what she describes as 'almost like a Cubist abstraction of the pattern', although O'Donnell + Tuomey have not used colour as surface patterning in that way.

Social housing at Galbally, Co. Limerick, Ireland,
O'Donnell + Tuomey Architects (2002)

Unité in Berlin,
Le Corbusier (1959)

Irish pavilion in the courtyard of the
Irish Museum of Modern Art, Dublin,
O'Donnell + Tuomey Architects (1991)

Meaning and association

It is often the case that the traveller returning to his or her own country sees the familiar with new eyes. Tuomey uses the term 'strangely familiar' to describe an aspiration for their work, but equally it could be used to refer to their own rediscovery of the traditions of architecture:[13] 'We would hope that our buildings would feel strangely familiar in the places where they are sited, and to the people who live with them. They have a compressed quality that comes from concentrated thought.'

The stark forms of Irish tower houses and farmhouses are more basic and unadorned compared to the poetic idyll of the rose-covered cottages of the English countryside. The 'little red shed', as the Irish Pavilion (1991) was known, derived its colour from the typical corrugated iron agricultural buildings of the Irish countryside. Seeing them again after a period of years away had a profound influence on O'Donnell + Tuomey. Such simple buildings are common in many societies and evoke strong associations. The red iron colour is 'naturalized through tradition of use and meaning', and was used in the pavilion celebrating the work of the painter Brian McGuire.[14] The spatial configuration of the small pavilion was crafted in response to psychological themes of isolation and of love, addressed by the artist in a series of paintings of prisoners. The small building had a sense of familiarity and memory, taken from the rural landscape, yet transplanted to the courtyard of the Irish Museum of Modern Art it was a shocking alien. McGuire was initially against the red colour and its connotation of the farm shed.[15] For him, the corrugated iron was more evocative of a border control post and should have been a dirtier, darker colour. They finally reached agreement using a secondary association – that of a Trojan horse, a container of unexpected, powerful paintings, set within a seemingly innocent shell. Although some visitors found it too much of a contrast, in the context of the formal courtyard of the museum, many mourned its loss when it was dismantled.

The red colour reappeared in the competition paintings for the first significant building in the landscape – the Blackwood Golf Centre (1994). Here, the intent was to subvert the normal associations of 'golf club' as this was to be an open public facility rather than a closed gentleman's club. The red-pigmented render used for the main segments of the Centre is used to make a visual connection to memories of farm buildings in Ireland, but it also resonates with the architects' love of Italy. The actual material used is a through-coloured pigmented render, which was imported from Italy, thus making both a symbolic and actual link.

The Howth House, for a private client by the sea in County Dublin, has emerged as a collaboration between client and architects. The client, a psychiatrist, was extremely interested in the way in which colour changes the perception of the spaces. Some of the original colours have since been changed, but only in consultation with the architects, who are also intrigued by the powerful effects of small changes.

Watercolour study for the Blackwood Golf Centre, O'Donnell + Tuomey Architects (1994)

The Howth House acts like a form of telescope, the plan form being conceived to heighten external views and frame the sunlit landscape to the north. Externally, the brickwork is coated with a grey-pigmented limewash, giving a neutral, subdued appearance. Internally, colour is used to emphasise the continuity of the free-flowing spaces, with an earthy aubergine colour used to give unity to the main walls, but with strong colours highlighting the exposed edges of openings cutting across the walls. Instead of painting individual rooms blue, or yellow, the colour is intended to reinforce the free plan and the spatial nature of the walls. The client has already repainted the walls several times, changing the colours to experiment with the reading of the space, in what O'Donnell describes as 'an ongoing conversation'. Unusually, the client wanted some spaces to be deliberately dark, and to have contrast with the more conventionally lighter spaces. Balancing functionality with psychological satisfaction seems to be an aim of both client and architects. The client completely understood the conceptual idea of the central hall as representing an outdoor space, even though it is entirely internal. The oscillation in the meaning of interior and exterior space is a thread which runs through a number of O'Donnell + Tuomey's projects, and is supported by their use of colour and materials.

Conceptually, O'Donnell + Tuomey consider the Howth House and the Glucksman Gallery in Cork to have an affinity for each other. Both play with curving walls, views and flowing space. The Glucksman Gallery uses predominately natural materials and white-painted walls to provide an appropriate setting for the art, with colour being considered temporal, applied by the use of chromatic light.

Seriality in the palette

O'DONNELL + TUOMEY have a tight palette of relatively few colours which are repeated in a series of works. Repeating colours from project to project supports the cohesion of their body of work. There is no sense from this practice that they plan to change their ethos or principal approach to making architecture. Although each individual project is, of course, unique, O'Donnell + Tuomey's approach to design mirrors their approach to colour, in that they hone their work as they progress, rather than feeling the need for constant invention in their methodology of design. The rigour of the architecture is upheld in the carefully considered limitation of the palette.

In this respect, they can be compared to Le Corbusier, whose restricted palette of 43 shades was, he considered, quite sufficient. Although O'Donnell + Tuomey have a copy of the Le Corbusier colour palette, and will refer to it from time to time, they have not found the specific shades to be of use, nor the prescribed combinations suggested by the use of the cardboard viewing frame provided with Le Corbusier's *Salubra* wallpaper range. They are, however, more interested in how Le Corbusier made use of the spatial effects of colour, such as his experimentation with achieving depth by the use of colours that appear to recede, and others that punch forward.

O'Donnell + Tuomey choose their colours to align with their aspiration to ground their work in the earth, site and construction. It is an architecture that shies away from artifice towards nature, and their choice of pigments and colours enhances and supports this. The appeal of the colours in the early Italian paintings was partly that the pigments were dug out of the earth, not squeezed from a tube, as would have been the case with artworks from later periods. It was this essentially natural quality that seems to have been particularly engaging to them, even if it was derived at the time from an impulsive affinity with particular compositions and colour combinations in Italian paintings.

The Irish Film Centre in Dublin was O'Donnell + Tuomey's first major project. Conceived as a courtyard, the key atrium space is clearly intended to be read as external, even though it is carved out of the centre of the urban block. Colour and materials are essential devices in the success of this inversion of inside space to outside atmosphere. This project exemplifies their first steps in defining the practice's use of colour. Instinctively, they sought to use natural pigment, which was sourced from L. Cornelissen & Son, an artist's supply company in London. The pigment arrived in great sacks of pure powder, reminiscent of the heaps of intense colour arranged on the floors of art galleries by the artist Anish Kapoor. The render was mixed on site as a simple sand/cement mix, incorporating the pigment, and applied to the interior atrium elevation in a single day. The air was yellow with the ochre.

Opposite the ochre wall, a small cubic form protrudes into the courtyard. Originally, this was finished in a strong cobalt blue applied in the same way as the ochre, although, in this case, unsuccessfully. Surprisingly, the blue pigment migrated from the surface into the substrate overnight, and the render had to be reapplied. Although the architects felt it was beautiful, the client was not as convinced by the uneven blue appearance. The adhesive from posters further damaged the wall, and the natural pigment has recently been coated with a clear red gloss, which is shockingly out of character with the original, gentle hues.

In their Ranelagh Primary School (1997, extended 2007), simple painted blockwork walls order the space, and different coloured doors imply a codification of use, which helps to make the building easily navigable. At the time of the design, O'Donnell + Tuomey looked at the work of Eileen Gray (1878–1976), the Irish furniture and interior designer who used blue, red, black and, occasionally, yellow in her work. Referring to the work of the South American architect Luis Baragán (1902–88), O'Donnell notes that they tend to test colour as he would, by painting samples onto large cards. The cards represent the office's tight palette, and have been taken out and reused on subsequent projects. The blue from the Ranelagh School, for instance, was subsequently used in their An Gaeláras Cultural Centre for Irish language, arts and culture in Derry (2009).[16] In both cases, the predominant building materials are unfinished; brick, concrete and zinc at Ranelagh and high-quality in-situ concrete in Derry. Colour is therefore used sparingly, and to give significance.

facing page
Irish Film Institute, Temple Bar, Dublin,
O'Donnell + Tuomey Architects
(1992)

SWEETS TICKETS INFORMATION BOOKSHOP

Ranelagh Multi-denominational School, Dublin,
O'Donnell + Tuomey Architects (1997)

New primary schools in Ireland were not, at the time, permitted to have plastered walls, in order to be both robust and economical. Most use simple materials, such as fair-faced concrete blockwork. O'Donnell + Tuomey complied, but used paint to give character to the blockwork walls and resin to give a feeling of depth to the floor. Ranelagh adopts the Fra Angelico palette of earthy red, yellow ochre, paleish green and a purply light blue. Although the light in Italy is entirely different, the original paintings, O'Donnell notes, were seen in dark interiors. In addition, classroom walls are painted white to allow the colour of the children's artwork to enliven the space. Where colours have been used, they are deliberately not strong primaries (which some will associate with primary schools), but have a more muted, milky quality.

Classrooms use the earthy red for key elements, such as built-in furniture in the classrooms, against the white walls and mustard yellow doors. The doors in the corridors are blue, whereas doors to the staff rooms are green. The colours have an underlying rationale, but are also placed by instinct, for example, using red adjacent to green. The use of contrasting colours gives a sense of harmony, which is directly influenced by the balance of similar pigments observed in the composition of the Italian paintings, 'They have a real sort of zing, but the opacity neuters them'. Part of O'Donnell + Tuomey's aim is to establish a character that is warm, yet has variety and a degree of complexity.

The dark grey 'RAL' colour,[17] which was used on the steelwork of their very first building, the Irish Film Centre, has been reused in Derry alongside their deep terracotta red from the Sean O'Casey Community Centre; as O'Donnell notes:

> maybe it's just the modern equivalent of Ford saying his cars could be any colour as long as they are black. You just get colours, and then they represent the concept of that colour … When you say blue, you obviously mean that colour, that's what blue is.

O'Donnell + Tuomey are concerned with the social and political meanings of art and architecture, colour is present to serve a purpose, to support conceptual associations. That O'Donnell + Tuomey's colour palette originated in the earthy natural pigments of Italian paintings should not therefore been seen as surprising. They aim to ground their buildings so firmly into the earth that they seem always to have been there. They dig into the site literally, and aim to do so metaphorically also. Their use of natural pigmented renders, limewash and a limited number of colours is consistent and deliberately timeless.

49
Hans-Fischer-Str.

– 5 –
Who's afraid of red, yellow and blue?
Eric Wiesner and Otto Steidle

Artist/architects

PAINTING and architecture have always enjoyed a symbiotic relationship. Art acts as stimulant, provocateur, a source of conceptual thinking and physical evidence. Art practice, and in particular painting, offers architects an exploratory medium; it is immediate, autonomous and provides a means to experiment, without the restrictions of time, budget and functional constraints. There are numerous examples of artist/architects, from Michelangelo, Raphael, and Karl-Friedrich Schinkel, to Charles Rennie Mackintosh, Le Corbusier, Bruno Taut and Zaha Hadid. Some use painting as a preparatory and experimental field, as part of a conceptual process. Architecture may eventually emerge, clarified and strengthened in intellectual rigour or grounded in intuition. One can see this in the work of British architect Will Alsop, who uses painting as a means of exploring ideas while deliberately trying to excise any sense of meaning. His paintings are intensely colourful, as are the buildings that eventually emerge, yet there is not necessarily any direct link between them. Alsop has a longstanding collaboration with the artist Bruce McLean, and while McLean's artistic work may be a final output, Alsop's own paintings are seen only partly as a means to an end. They are abstract painted interpretations of conversations, programmes and public consultations. He notes:

> one of the reasons for painting is that you are not really in control of what you are doing – and that interests me a lot … It is not always necessary to know what one is enquiring into. It is the work that dictates a direction.[1]

facing page
Wohnquartier am Theresienpark, Munich (housing quarter), Steidle Architekten (2007)

Painting, Will Alsop

Other architects enjoy a more collaborative relationship with the artist, particularly when considering colour. Mark Wigley notes:

> while the painter and the architect are destined to meet on the surface they share, the meeting necessarily takes the form of a confrontation because of their different attitudes towards that surface. It is not that the architect provides the space and the painter provides the colored emphasis of that space. It is the color that provides the space.[2]

The German artist and sculptor Erich Wiesner, for example, has worked with a number of architects, most notably Otto Steidle and Günther Behnisch. It is increasingly common to find architects either deferring to an artist or colour consultant, or forming teams with artists, through which collaboration responsibility for the conceptual and specific use of colour is shared. Of the architects selected for this book, a number have collaborated with artists, specifically in relation to colour design – Caruso St John with Thomas Demand and with Lothar Götz, Gigon/Guyer with Adrian Schiess and Harald F. Müller, and AHMM with Charlotte Ingle, Martin Richman and Studio Myerscough.

Considering the Modernist housing estates or *Siedlungen* of Bruno Taut and Le Corbusier, the work of artists such as Bridget Riley, Donald Judd, Yves Klein, Mark Rothko and David Batchelor, and Steidle and Wiesner's housing projects in Turin, Ulm and Munich, this chapter will offer insights into the way in which artists deploy colour and their influence within architectural collaborations.

Heroic Modernism

ARTIST/ARCHITECTS, such as Le Corbusier and Bruno Taut, had a profound impact on the development of Modernist design in the heroic period of the 1920s. Their interest in colour stems from painters and from their own art practice.[3] Both worked between the two disciplines, and although Le Corbusier also collaborated with the artist Amédée Ozenfant in his writings until 1925, his *Polychromie architecturale* is entirely his own, and he continued to paint all his life, culminating in his 1955 publication *Le Poeme de L'angle Droit*.[4] The Modernist architects reassured themselves that it was safe to experiment with colour without it being seen as degenerately decorative, because the art to which they referred used colour in an abstract, rather than pictorial, manner. Flat surfaces or block colour could readily be transferred to walls, frames and planes. Painting and sculpture, therefore, had a direct influence on the way they thought about space and volume.

Bruno Taut visited Berlin as a young man and was, as noted by Iain Boyd Whyte in his book *Bruno Taut and the Architecture of Activism*, bewitched by the painter Arnold Böcklin. Whyte quotes Taut's letter to his brother back home: 'The poetry of the colours is indescribable. I'm searching so hard for words, but words fail. One can simply look … and keep silent'.[5] In contrast to his subjective – and as he himself confessed irrational – passion for Böcklin's paintings, Taut was impressed by an objective and rational sense of order, reason, rhythm and harmony that he observed in a department store by the architect Aldred Messel. At this point in his life, Taut wrote that he was torn between being an architect or a painter. Perhaps this is the root of his interest in the spatial, as well as the emotional, possibilities of colour.

As he started to practise, colour became intrinsic to his architecture, and to his socially driven agenda. The community at Falkenberg (1912–14) employed repetitive house types, consistent materials and scale, but used colour to give an appearance of diversity and to reinforce the reformist ambitions of a 'vigorous social mix'.[6] Harmonious colour was used as an analogy for harmonious living between mixed social classes and groups. One modest ambition of a manifesto *Die Erde eine gute Wohnung* (1919) (*The Earth a Good Dwelling*) published by Bruno Taut, in support of the experimental housing projects, was that people should not have to transform and modify their behaviour to suit buildings, but rather that architecture should correspond to human needs.[7] Perhaps the best-known of Taut's housing projects is the *Waldsiedlung Onkel Tom's Hütte* (1926–31). Recently restored to the original colours, but with the benefit of mature landscaping,

Waldsiedlung Onkel Tom's Hütte, Bruno Taut (1926–31)

Bruno Taut's own house in Dahlewitz, Berlin (1927, since partially restored)

the homely atmosphere of the place is keenly felt. It seems to support living without prescription, effortlessly accommodating the immense social change that has taken place since it was built. Berlin has recently been awarded Europa Nostra status for the *Siedlungen* built across the city in the 1920s. The simple urban planning of streets and gardens, fronts and backs gives a clear spatial order. The repetition of house types brings rhythm and the use of colour both unites terraces and contrasts streets, giving individual identity.

Taut's choice of colours is occasionally so strong that he could be accused of breaking his own proclamation, that the architecture should not dominate the utility of the space. It is hard to see how it is possible to live with them. His own house at Dahlewitz (1926–27) combines red, yellow and blue in an extreme fashion. Indeed, so shocking were his combinations of colour that Le Corbusier, strolling through the *Weissenhofsiedlung* in Stuttgart (1927), is reported to have exclaimed 'My God, Taut is colour blind!'[8] Taut, it seems, had no lack of confidence and saw no need to collaborate with an artist.

Arthur Rüegg notes that *Onkel Tom's Hütte* was built at the same time as Le Corbusier's housing at Pessac (1924–26). Whereas Taut used traditional construction methods, Pessac used systemized production methods to mass produce workers' housing, applying principles of flexible space. Pessac was Le Corbusier's first project in which colour was used extensively and deliberately to create specific spatial effects. During this period, Le Corbusier's preference was for colour solidity and a sense of unity. Although his understanding of colour was, at this point, limited, he noted some basic observations; for example, that a change of colour at a corner internally breaks up the volume of the room, and that dark (light-absorbing) colours dissolve the physical appearance when used externally.[9] Pessac is an experiment in proximity, in foreground and background, the spatial effect of colour and in the composition of elements – all aspects of painting.

The Pessac houses are painted pale green, blue, red ochre and white. The blue was used to expand space by appearing to recede from the viewer, the red to establish fixed points within the urban configuration, the green to interact with the landscape and the white as a calibrator against which the other colours are enhanced.[10] The project embodied a well-documented paradox. The dwellings were intended to accommodate change easily, but the inhabitants chose to add pitched roofs and subdivide open-plan spaces, subverting much of the architect's vision. Ironically, the Pessac houses are now being stripped back to their pure state by the next generation of owners.

Le Corbusier's showpiece houses at the Stuttgart *Weissenhofsiedlung* (1927) use colour as both a rarifier of architectural elements and to induce spatial effects. The pale sky blue of the terrace walls increases the openness of the roof terrace. The earthy-brown vertical plane of the spine wall anchors the building into its site. Some texts acknowledge the contribution of site architect Alfred Roth in relation to colour.[11] Others suggest that the colour was very much that of Le Corbusier. The Weissenhof buildings are examples of polychromy as part of his *promenade architecturale*. The surfaces are painted pale yellow, peach and grey, and a muddy brown – colours used in his abstract paintings. Internally, it was not wholly successful, according to Theo van Doesburg, who scathingly compared the dark brown painted corridors at Stuttgart to the experiences of the First World War trenches. He also notes that there was too much emphasis on the aesthetic composition of 'purist-pittoresque' colour, rendering the spaces uninhabitable.[12]

Arthur Rüegg suggests that, in many ways, Taut and Le Corbusier had similar approaches to colour. Where they can be seen to differ is that Taut was less restrictive and was open to influence by context, sunlight and shadow. Although both were aware that the choice of colour has the effect of enlarging space or reducing distance between houses, Taut equated colour directly with light.[13] He favoured bright colours adjacent to each other internally and in areas of shadow externally, while Le Corbusier would tend to restrict bright colours, against a muted background. Le Corbusier advocated a highly restricted palette, whereas Taut was altogether less doctrinaire, less systematic.

'Colour Field' and the emancipation of colour in the 1950s and 1960s

DONALD JUDD'S work is grounded in his exploration of colour and space. Having observed that the 'Colour Field' painters of the 1950s and 1960s, such as Barnett Newman, Frank Stella, Morris Lewis and Mark Rothko, had taken flat colour almost as far as he thought it could go, he moved into three dimensions and simple forms. The spaces between objects became part of the compositions. His studies as part of the *Progressions* series appear to relate the intervals between objects mathematically. Yet as Nicholas Serota notes:

> I think appreciating Judd's work is very much about feeling rather than rational thought ... most people when they are looking at Judd's work assume that it's all worked out to the millimetre and, in a certain sense, it is very precise but it's not that mathematical, it's a combination as he put it of intuition and feeling and rational thought trying to discover in a certain sense the unknowable within the rational.[14]

Judd has been associated with cadmium red but, for him, the colour itself was not of prime importance, but rather how the colour affected the perception of an object to which it was applied. Black would blur the edges of an object, white would diminish it, but red, by contrast, gave sharp definition and a quality of light surrounding the object.[15] In one piece, a rectangular copper vessel is painted with cadmium red only on the internal base. Reflections in the copper internal faces appear to fill the container with red colour. Is this sense of the purity of colour for its own sake, free from association and function, something that artists can contribute to an architectural project?[16]

Untitled, Donald Judd
(1990)

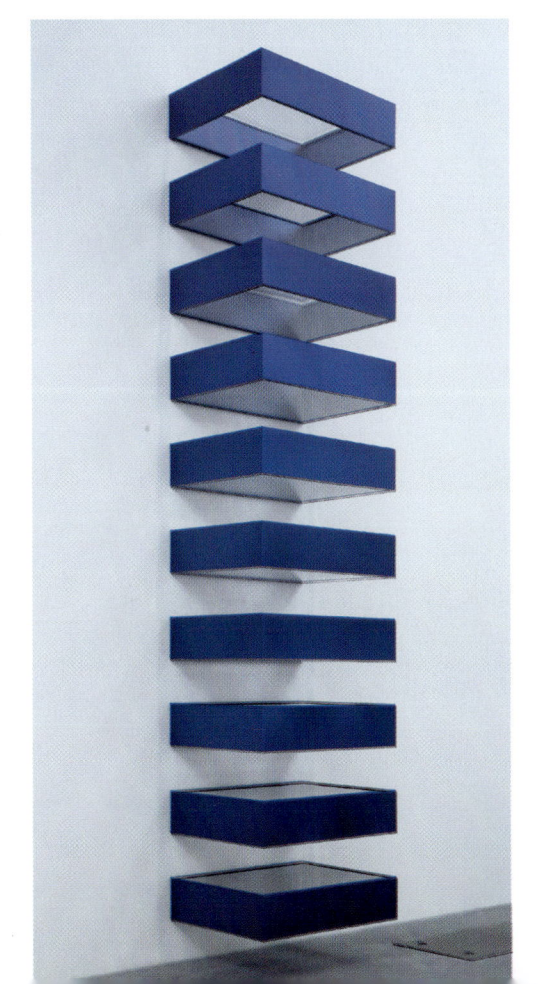

Andante 1,
Bridget Riley (1980)

The artist Bridget Riley's work employs beats of irregularly spaced, but often repetitive, coloured stripes or shapes. She considers that, 'Painters see only pure colour not as the colour of something'.[17] Colours are employed for what they do. Robert Kudielka in his essay on Riley's work notes that, 'from the sheer wall of the paintings there always echoes back the same laconic answer: nothing else, there is nothing else behind it. Bridget Riley's art contains – in the strict sense of the word – no content.'[18]

Around 1967, Riley switched from black and white Op Art paintings of the 1960s to vivid colour, a painful process that took over two years of trial and dissatisfaction. Her paintings worked with colour instinctively at first and then, only after some years, did she begin to plot the optical relationships of the colours. Only in retrospect could she reflect on the framework within which she articulates the 'colour event' of the painting. Technically, the paintings combine both the effects of simultaneous contrast and the fusion of contiguous colours. In *Andante 1* (1980) only five colours are used, but fugitives appear between them as they interact in the eye, making a more complex reading of a simple structure.[19] She is wary, however, of too much rationalization of the physics, because as soon as the artist becomes involved in the messy reality of pigment 'the scientific purity is lost'. It is not just the imprecision of the medium, it is also the unreliability of the colour itself, subject to light conditions, viewing distances and the perception of the observer, that fluctuates. The very plasticity of the medium has become the essence of her work. Riley is one of a number of artists who are regularly cited by contemporary architects as being influential on their work. In both Riley and Judd we see structure, rhythm, and space, all directly applicable to architecture.

Authority, originality and the creative impulse

NERVOUSNESS, or even prejudice, infiltrates art as much as it does architecture. Colour is thought to be dangerous, risky, and unreliable. It is little wonder that architects search for easy-to-apply rules and look to art to inspire and teach. In any creative pursuit, strength comes from the moment of expression, from an instinctive emotional response. Matisse's intense red painting, *The Red Studio*, for instance, depicted his feelings towards the space, which in reality was almost entirely grey. Its expression is wholly visceral. Barnett Newman's painting *Who's Afraid of Red, Yellow and Blue?* (1966) queries the subjugation of colour within art. The 1960s was generally a confident period with regard to colour.[20] Such confidence comes in cycles and is countered by periods of 'chromophobia', as so clearly identified by David Batchelor.[21] If artists can lack confidence in the use of colour, what hope is there for architects? What more can we learn from the artist in relation to colour use, selection and juxtaposition, and what happens when artist and architect collaborate?

The abdication of authority to – or the sharing with – a collaborator can be reassuring. Designing architecture involves many thousands of small daily decisions. There are areas in which the architect feels totally in control, and there are other areas where architects may deliberately invite an element of chance. Colour can so radically alter the perception of space, possibly detrimentally, that architects find themselves in uncertain territory. John Ruskin, the nineteenth-century art critic, associated colour with contingency: 'The artist who sacrifices a truth of form in the pursuit of a truth of colour; sacrifices what is definite to what is uncertain, and what is essential to what is accidental'.[22]

Society, as the painter Mark Rothko observed, can be unfriendly or even hostile to the activity of an artist, not least, as the work can be unfamiliar. But defamiliarization is a key role of art, as Viktor Shklovsky surmised:

> Art exists that one may recover the sensation of life; it exists to make one feel things, to make the stone *stony*. The purpose of art is to impart the sensation of things as they are perceived, and not as they are known. The technique of art is to make objects 'unfamiliar', to make forms difficult, to increase the difficulty and length of perception because the process of perception is an aesthetic end in itself and must be prolonged.[23]

Rothko considers his paintings as 'an unknown adventure in an unknown place' and suggests that the artist has to accept the lack of security inherent in the process in order to be truly liberated.[24] This romantic ideology is also present within any original design activity, for 'the creative impulse resides with genius, the illusive, innate, sometimes childlike, site of creativity in each of us. In this tradition genius (essence) is closely aligned with chance (the accidental). Both absolve us from the necessity to explain.'[25]

In instances where the artist is an invited collaborator, both client and architect invest in a measure of uncertainty. The architect may assume a central position between artist and client and may act as a moderator. Johannes Ernst, partner in Otto Steidle Architekten, observes that this is sometimes simply impossible.[26] 'Colour requires strong decisions and one needs a strong project and a strong client if it is to be productive. Occasionally, the colour takes on an autonomous position – the artist coming in one direction and the architect another'.[27] To collaborate successfully, the architect must accept the validity of the artist's position. Some architects, however, simply would not countenance such dilution of control; viewing the artist, as well as colour itself, with suspicion as an agency 'dangerous to architectonic form and the authority of the architect'.[28]

Architects can, however, be nervous about, or mistrust, their own ability or understanding. This may be partly due to a lack of training or emphasis in contemporary architectural education, or simply because some architects associate artists with a more highly tuned sensibility and understanding of colour. A collaboration with an artist may be highly productive, relaxing control, bringing a fresh impulse that, particularly if considered early in the design process, can be successfully synthesized. Deferring to the authority of an artist can also provide a vehicle through which colour can be more easily discussed with the client. Any lack of rational explanation can be countenanced by both parties as there is a general, if conventional, expectation that artists are motivated by emotional responses. If, as is evident in the collaboration between Berlin artist and sculptor Erich Wiesner and Steidle Architekten, the architect is open-minded to the interpretation and enrichment that an independent mind can bring to a project, the authority becomes shared and the colour is totally absorbed into the architecture.

Wiesner and Steidle: artist/architect collaboration

'As soon as colour radiates into space you know whether it's water or Schnapps.' [29]

Fear of colour is in no way apparent in the work of Erich Wiesner. Aware of its potential impact however, he compares colour to a small glass of schnapps – apparently harmless, but, even in small quantities, very powerful. Johannes Ernst warns:

> If you work with an artist, you work with an artist. … The purity in art, of course, is always somehow a guideline. This is something completely different from studying theory and practice. This is why working artists are not art historians. They are artists. The moment of creation, and the moment of design, is completely irrational. It's completely unexplainable [sic]. It's completely indefinable … the artist understands the emotion. This doesn't have anything to do with theory.[30]

Otto Steidle's office has had a close working relationship with Erich Wiesner for many years, but they were thrown together by chance for a 'Kunst am Bau' [Art in Construction] project for the International Center for Academic Exchange (IBZ), designed by Steidle and built in 1983. The initial relationship appears to have been quite difficult, perhaps because Steidle had studied art for two years and considered himself in the artist/architect mould, and did not feel any need to collaborate. He was uncomfortable when asked to work with Wiesner, who was then predominately known as a sculptor. Johannes Ernst commented that the two men found solace in a shared passion for *Pflaumenkuchen* (plum cake), a deeply-held sensuous memory that seems to have cemented the relationship. Thus began an enduring partnership. Looking back, Ernst suggests that the most important element of the collaboration is the freshness, the moments of complete surprise – on the part of both artist and architect – even in a longstanding relationship. For a large waterfront project in Hamburg, Steidle's office introduced a sharp, angular plan form unlike their normal rectangular geometry. Wiesner, astonished by the form, reacted by using colour to soften the sharpness of the angles, not by the choice of colour itself, but by pixelating the facades with coloured glazed bricks, to create fractal patterns across the angular facades. The result is a clear break from the office tradition, but one that has spawned a number of further investigations of surface. Hermeneutical theory suggests that the design process is imbued throughout with interpretation. This is not to suggest, of course, that artists are without reason, any more than one would suggest that architects are devoid of original creative impulses. Rather, that the continuous demand for economic justification and the high levels of responsibility that mark the practice of architecture make the subjectivity and complexity of colour tricky to defend.

Rhythm and blues

OTTO STEIDLE'S architecture was rooted in the humanism of his generation of architects that included Herman Hertzberger, Ralph Erskine, Lucien Kroll and Giancarlo de Carlo. His early works, such as his own office in Genterstraße, Munich (1972), are demonstrably part of structuralist investigations of flexibility in occupation offered by prefabricated precast concrete frames. The open structure was intended to accommodate change and re-interpretation by the occupier, and has demonstrably succeeded. Steidle sought to blur the distinction between a building for living and a building for working. The skeletal, prefabricated and repetitive framework is now barely recognizable as the landscaping has matured. Steidle's architecture, not only in the use of colour, has some similarities to that of Bruno Taut. Both are known primarily for their housing projects, which draw on the central European tradition of colour used externally, and both place the user at the centre of a socially derived agenda.

The first major project with Erich Wiesner was the University of Ulm West Faculty of Engineering Science (1988–92). The 400-metre-long megastructure creates a promenade through the wooded hillside above Ulm. Steidle sought to undermine the stereotype of scientists as serious, grey and rational, by making a creative environment more typically associated with the arts – spontaneous, diverse and colourful.[31] Wiesner was the obvious conspirator, using colour to provoke, challenge and energize. He has likened the iterations of colour and the rhythm of the structural elements to a diagram of the *Fugue in C Minor* by Bach. The essential quality of a musical fugue is one of themes, repetitions and variations, and the notation is a wonderful example of visual densification. The timber frame elements at Ulm set the beat and the spatial relationships of the segmentation, as well as giving a rational, regular rhythm for the facades of the main promenade. His interpretation of Steidle's desire for spontaneity was through colour. Whereas the regular structure retains a sense of authority and order appropriate to a university, Wiesner's colour design introduces secondary, irregular, pulses across the facades.

The process of colour selection for the project at Ulm University, unlike most collaborations between Steidle and Wiesner, was relatively quick, but still meticulous. The time and effort invested in honing colours to very specific shades is common to most of the architects interviewed. This is an intense process, and is one of the most obvious distinctions between architects who use colour as

University of Ulm West, Otto Steidle/Steidle Architekten, colour design by Erich Wiesner (1992)

Erich Wiesner's Studio, Berlin

an essential integral part of their architecture, and those for whom colour is a necessary evil, chosen rapidly and, perhaps, naively. Sometimes such finessing of tone is simply not possible within the economics of construction projects and the colour choice has to be less precise. In one project, the contractors could not accept Wiesner's initial proposal that included six or seven different shades of white. 'White is white', he was advised, and was forced to accept the economic imperative.[32]

At Ulm, he started with 48 colours, these were edited to 24 and finally to 15. Wiesner seems to know each colour intimately, to the extent that he notes poetically:

> From 48 to 15
> or
> On finding the colours – more realistically;
> On being found by the colours
> ... They have no prejudice, no ambition, no ego.[33]

Wiesner thinks of colour with the same conviction as Riley, as something pure. His open-mindedness when it comes to choosing specific colours is clearly valued by the architects with whom he collaborates. He will approach colour 'like a beginner', but acknowledges that one has colours that one trusts, and will repeat them – a certain blue, or a certain yellow. He tries not to think of associations with colours. For the St Benno Gymnasium, Dresden (1995), this time with architect Günter Behnisch, the local *Dombaumeister* advised that they should not use a particular green or this blue because they were German Democratic Republic (GDR) colours, but Wiesner was not concerned by these restrictions.[34] Although open-minded to any colour, this sense of trust in particular shades is common. Architects O'Donnell + Tuomey restrict the available palette to a limited range, as did

Le Corbusier in his *Claviers de Couleur*. Wiesner considers yellow-greens to be the most dangerous and tends to avoid them, favouring warm orange-yellows instead. His starting point is often red, yellow and blue. Colours one might associate with Mondrian, Kandinsky or Paul Klee, if used in their primary state, are overpowering in the large scale of architecture. Gerrit Rietveld was aware of this and restricted their use to small, floating planes in both his architecture and furniture. Wiesner tends to mute the tones, resulting in a palette that is still strong, but slightly subdued. Green, when introduced, is pale, almost opal green. Reds are either poppy-red or veer towards terracotta, but he continues to have a fascination with blue despite it being perhaps the most difficult and controversial colour to use architecturally.

The search for the perfect blue has perplexed Wiesner almost as much as it did the artist Yves Klein, who resorted to chemistry to patent his 'International Yves Klein Blue' in 1957.[35] Wiesner laments the range of blues available, and has been known to stand over the paint manufacturer like an alchemist, subtly modifying the mix. Keim mineral paints are frequently used for this reason, as the company will mix any colour and retain precise records of the mix in their archives. The lack of good, readily-available blues may partially account for visual mistakes by less experienced colour designers where an inappropriate blue has been used. For Steidle's housing project for a site at Bruderstrasse, opposite the Haus der Kunst in Munich (1997), Wiesner introduced a muted version of Yves Klein's ultramarine on a tall central block. He considers that 'no other color promotes the fading of the visible as much as blue'. Johannes Ernst observes that blue is such a key colour in art:

Housing at Bruderstrasse, Munich, Steidle Architekten, colour design by Erich Wiesner (1997)

> it is the colour of the romantic period in painting. It's the colour of desire. A blue, of course, is completely associated with the hand of Yves Klein. So to make a blue building when blue is the colour of the sky, is often criticized. … a building can't be blue, they say, the sky is blue.[36]

At Freischützstrasse in Munich (1997–2001), a further collaboration between Wiesner and Steidle, the combination of simple terraces and intense external colour is immensely powerful in defining a series of urban rooms. For this project, Wiesner created a colour scheme that articulated the volumes in two contrasting colours to make them look less massive. The urban principles are similar to their project at Wohnquartier am Theresienpark, also in Munich (2007), staggering long and short rows to contain courts. The end gables facing the courts are the strongest hue, deep terracotta red with ultramarine blue main walls. He ignored any articulation of

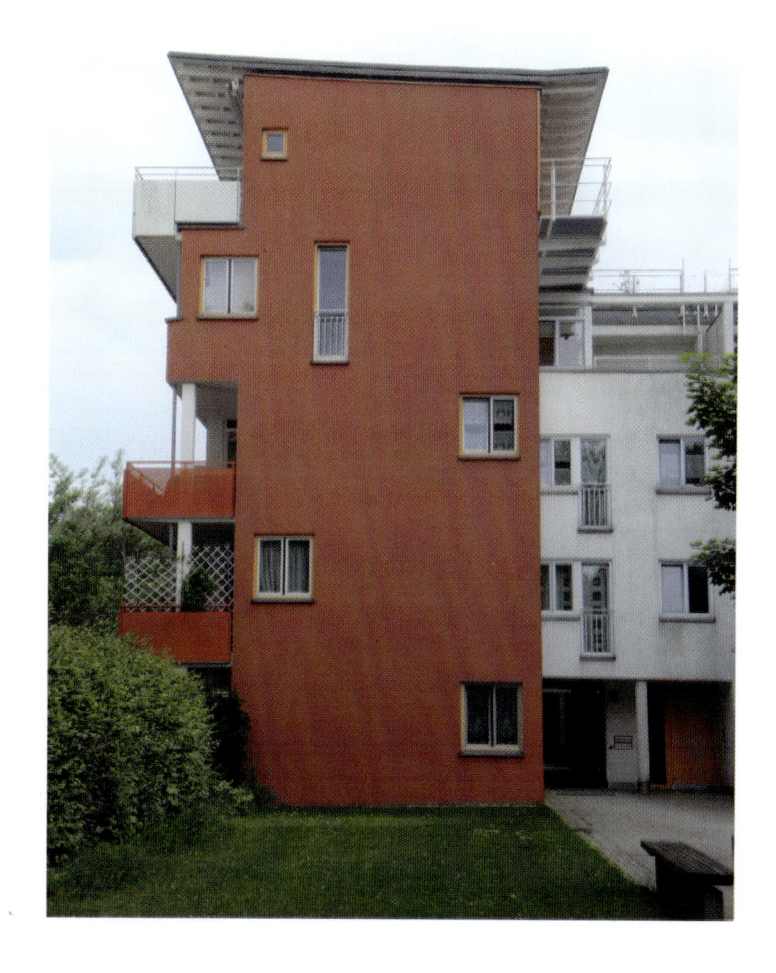

Housing at Freischütstrasse, Munich, Otto Steidle/Steidle Architekten, colour design by Erich Wiesner (2001)

change of materials and so balconies, rather than being expressed as an element, are coloured the same as the walls. His argument was that this calms the building and transforms it into 'powerful bodies of colour'. Perhaps it is Wiesner's background as sculptor, rather than artist, that makes him comfortable with the expression of three-dimensional space and colour. White is occasionally used on blocks to give an overall visual and experiential balance. The aim is to deliberately increase the complexity, variety and contrasts, despite the simple form. Oliver Hamm comments that the owners were initially sceptical of the strong colours, but this reservation has long since passed:

> No one is afraid of red, yellow and blue any longer ... colour in architecture – and especially domestic architecture – as used by Bruno Taut three-quarters of a century ago, but confidently and naturally, seems to have become socially acceptable again.[37]

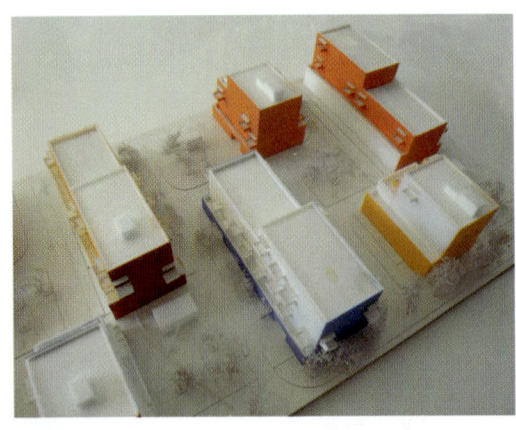

Study model (left) of housing (below) at Theresienpark, Munich, Steidle Architekten, colour design by Erich Wiesner (2007)

The rhythm of structural frame, or sub-rhythms of timber truss, joist and repetitive elements evident at both Ulm and Freischützstrasse is characteristic of Steidle's early work. In the office's more recent works, walls have become more homogenous and the colour more solid. The rhythms have developed from showing an overt expression of structure to investigating an interest in the relationships between buildings at an urban scale. In place of colour as play of surface, Wiesner developed colour as a volumetric device to infect space. The housing project Wohnquartier am Theresienpark, in Munich (2007), sets out a pattern of simple box-like objects, staggered in position to create and contain a series of urban spaces. In the model studies, surfaces are coloured to give character and identity, not just to the objects but, more particularly, to the spaces in between.

As in Donald Judd's *Progressions* series, the spaces between the objects are defined by material, space and colour. For Rudi Fuchs, director of the Stedelijk Museum, Amsterdam, curating an exhibition of the work of Donald Judd, 'it was because he wanted to concentrate on articulating color in space, that he preferred these simple forms; they created spatial clarity in a more precise way than complex forms which attract too much attention for themselves'.[38] Unlike Judd's work, the architecture must take account of function in addition to the physical experience of the spaces, and the changing perception as the moving eye threads between the blocks. Although the spatial

Olympic athletes' village, Turin,
Benedetto Camerana, colour design by
Erich Wiesner (2006)

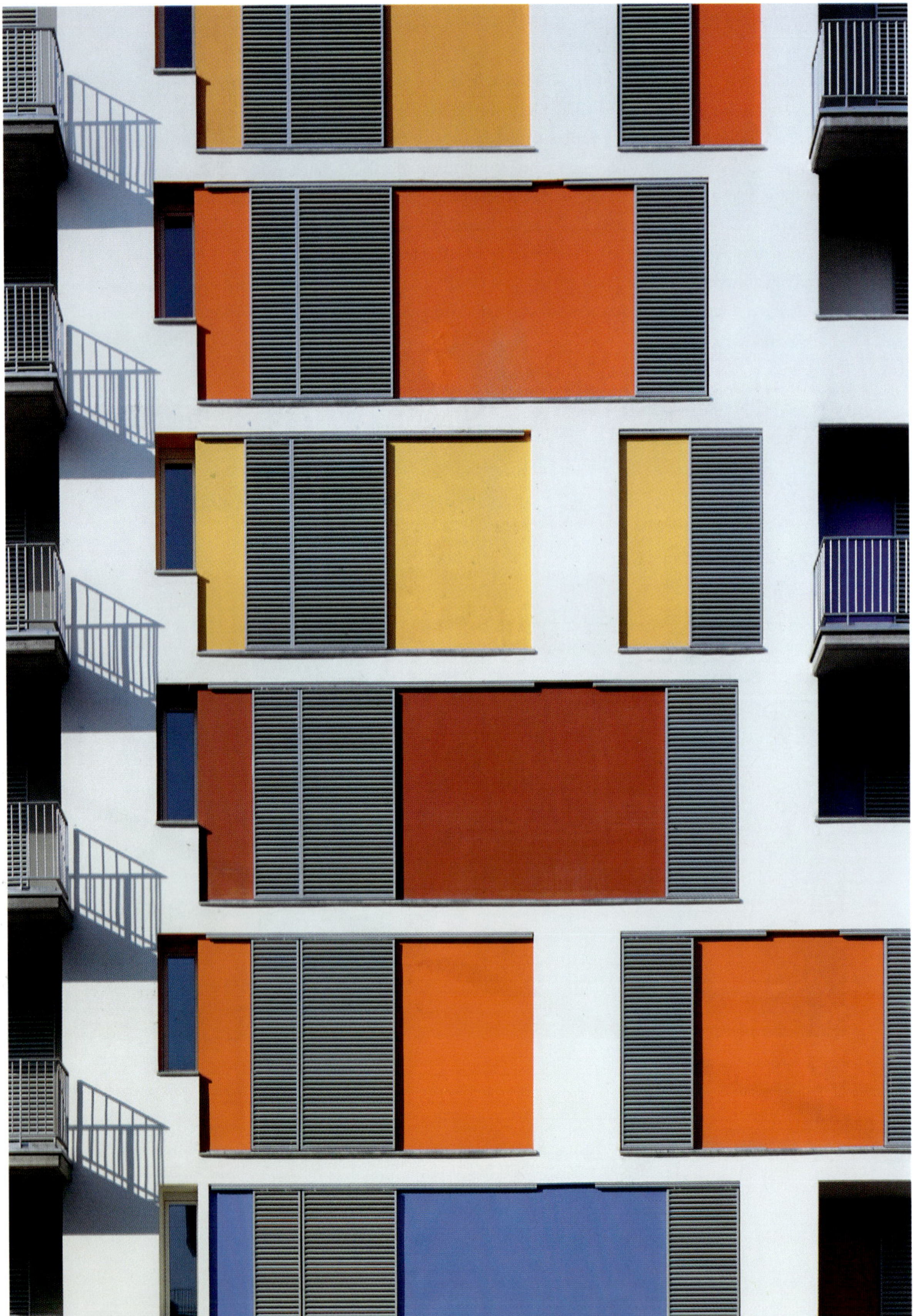

relationships are similar, and the blocks repeated, the colour completely changes the feeling of the space. The interaction of colour and space through choreographed sequential intervals is, however, reminiscent of Judd's compositions.

The athletes' Olympic village in Turin (2006), which the Italian architect Benedetto Camerana won in 2002 in competition with a number of European architects including Otto Steidle, is one of the best examples of the contribution of Erich Wiesner to architectural and urban colour design.[39] The overall masterplan deliberately puts an emphasis on pedestrian circulation in and around a series of blocks. Camerana subsequently invited Diener+Diener, Atelier Krischanitz, Ortner & Ortner, Hilmer & Sattler and Steidle Architekten as consultant architects to work with Wiesner. Steidle and Wiesner had, by then, executed a large number of projects together and had a clear working understanding. Camerana, however, was utterly shocked by Wiesner's interpretation of his project. He simply did not recognize his buildings.[40] Despite this reaction, the Italian boldly went ahead with the project, accepting the suggestions of the artist. Johannes Ernst, one of the directors at Steidle Architekten, is of the opinion that it is Camerana's best building, but notes that the architect has, apparently, not been willing to repeat the process.

Experimental echoes

M ODERNIST European *Siedlungen* of the early twentieth century held an expectation of experimentation in both architecture and in projections of future lifestyles. The tradition has re-emerged from time to time and, notably, in respect of the use of colour, in suburban Vienna. Otto Steidle was one of three architectural practices invited to design a contemporary *Siedlung*, of around 200 dwellings at Pilotengasse (1991). It was one of the first housing projects of recent years

Social housing at Pilotengasse, Vienna, Adolf Krischanitz Architect, colour design by Oskar Putz (1991)

facing page
Facade, Olympic athletes' village,
Turin, Benedetto Camerana,
colour design by Erich Wiesner (2006)

to make colour an integral part of the urban design. The project mimics the experimental Modernist *Siedlungen* of Stuttgart, Vienna and Berlin – part exhibition, part social experiment – expressive of a democratic community living together. The overall masterplan, designed by the Swiss architects Herzog & de Meuron, provided a strong formal layout of solid lines and gently curved voids that unify the estate. Colour divides each section, not only in the different hues, but also by the way in which each architect has employed the colour in support of their architectural and social expression. There is a sense of tension between these ideologies. It is not simply the colour in this project that is of interest, but what the colour does.[41]

In each case, the architects worked with an artist: Adolf Krischanitz and Otto Steidle with the Viennese artist Oscar Putz, and Herzog & de Meuron with the Swiss artist Helmut Federle.[42] The eastern outer line, by Krischanitz, gives emphasis to individual houses, set at right angles to the main blocks. They appear like a row of lightly tethered caravans facing inwards and are out of character with the terraced typology. These two-storey houses are designed as variations on a theme, but express a sense of individuality though the colour. Although contrasting with each other, the vibrancy of the artificial hues of yellow, turquoise and lilac then tie them in firmly as part of the neighbourhood. Putz drew most directly on his studies of the work of Bruno Taut, using strong contrasts of red and green on the outer curve of one linear row, as used in *Onkel Tom's Hütte*, marking each house as distinct from its neighbour. On the inner face of the same row, the colours sweep from red through orange to yellow and back again. As front and back are never seen together, the faces of the terrace take priority conceptually over the form. The irregular rhythm of the windows, then adds a random layer to the ordering.

The approach to colour by Federle with Herzog & de Meuron is altogether different. The architects, known for their highly eclectic output, have an uncomfortable relationship with colour, with some notable successes (such as the Laban Dance Centre in London's Docklands near Greenwich, London) and some failures (such as the cold dark blue of their Barcelona Forum). At Pilotengasse, they draw on natural mineral colours of earthy browns and greys. The intention was to be 'colourless', with the texture and tone generated directly from the sandy material. On the western side, which is most exposed to the weather, a coarser texture is used with a much finer grain on the inside face. Originally, the design included vertical stripes, not specifically to express the separation of one house from another, but as a reference to the pragmatic need for breaks between applications of plaster. In the end, colour was applied to the surface, as the quality of the natural render was not sufficiently uniform. The proposed stripes soon became conceptually unjustifiable. The result is homogenous, and expresses the unity of the urban terrace. The rhythm of punched windows and extruded forms therefore dominates over the surface, and the subdued palette can be seen as a natural antidote or counterbalance.

Sitting centrally, both physically and philosophically in respect of the colour usage, are the three rows designed by Steidle + Partner. The unity of the terraces is expressed by a consistent warm blue tone, with strong yellow, blue and turquoise restricted to recesses. In this case, colour is used in support of architectural form, solid and void as opposed to surface. The pale blue walls recede from the observer, apparently opening up the central, most significant space. Although Oskar Putz worked with both Krischanitz and Steidle on this project, the articulation of social and formal cohesion by Steidle and of individuality by Krischanitz appears to have been strongly influenced by the aspirations of each architect.

The impact of artists is extremely evident throughout this book. In some cases this is a product of a general culturally literacy on the part of architects. Occasionally, architects may source particular combinations of colour shades directly from paintings. Increasingly, however, collaborations between artists and architects are proving successful. The longstanding and symbiotic relationship between Otto Steidle and Erich Wiesner is one such example. For Wiesner, architecture provides a radical shift in scale and a significant public platform. Architecture amplifies the interaction between the object in space and the perception of the moving body moving through space. In return, artists provide an understanding of the essential plasticity of colour, and an assurance that uncertainty and emotion are justifiable, desirable constituents of architecture. Le Corbusier and Amédée Ozenfant considered the work of art as 'an artificial object, which permits the creator to place the spectator in the state he wishes'.[43] Colour used in architectural design can be potent, by immersing the whole body in a spatial, perceptual experience.

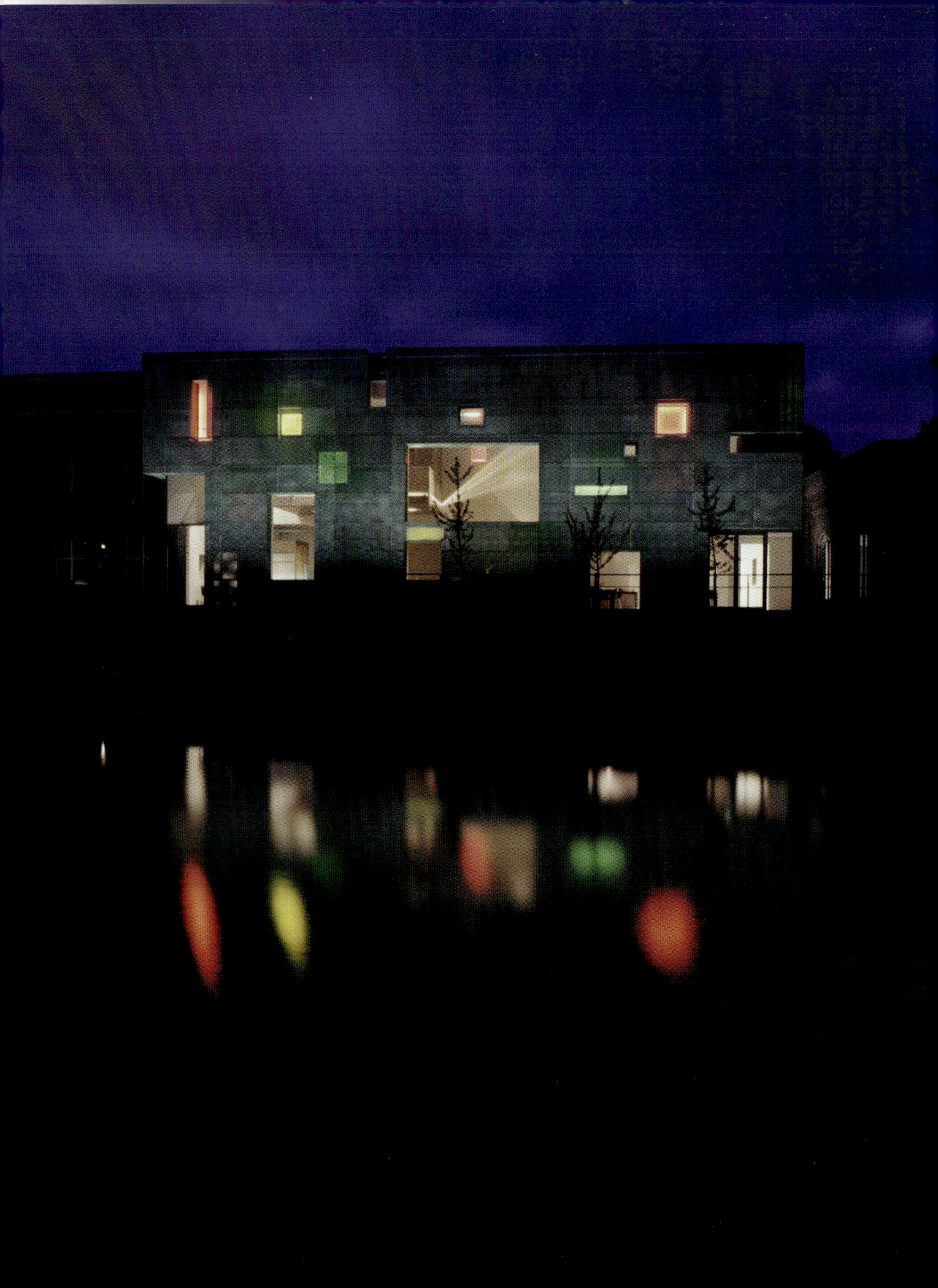

– 6 –
Place, space, colour and light: Steven Holl

*'Light is the most noble of natural phenomena,
the least material, the closest appropriation to pure form ...
light is the principle of order and value.'*[1]

ALTHOUGH pigment and pigmented materials form the focus of this book, the effect of light conditions on the perception of space and material surface is profound, and offers a further dimension of opportunity for colour design. Colour generated through the play of light is never static and has the capacity to be used as an instrument to tune and transform architectural space. Although coloured glass has been used for centuries, the symbolism embodied in window design was the primary intent, and the interaction of the cast colours across interior surfaces largely a secondary effect. In considering first the metaphysical properties of colour and light and the interaction between them, which can be seen as instrumental to the synergic design of architectural space, and second the role of architectural colour in contributing to a specific sense of place, the chapter is contextualized by the work of the American architect Steven Holl.

Colour and light

HOLL first explored the use of tinted and reflected light in his design for the offices of D.E. Shaw, a centre for financial trading in New York (1992). Colour was applied as paint, but to surfaces hidden from view at the base or back of notches set into wall surfaces. His intention was to stimulate curiosity and give a sense of the intangible. Moving through the space, the reflected

facing page
Reflected light as pools of colour in the canal adjacent to the offices
on Sarphatistraat, Amsterdam, Steven Holl Architects (2000)

Chapel of St Ignatius, Seattle,
Steven Holl Architects (1997)

light is sufficiently strong to drift across space and surface and across one's body, engaging the users as participants in the optical illusion. The paint colours used are almost fluorescent – an intense blue, orange and green – chosen not to be seen directly, but only as devices to tint light.

The Chapel of St Ignatius in Seattle (1997) is undoubtedly one of Steven Holl's key projects that employs colour as an integral part of the design. Simple in conception, the project synthesizes architectural precedents, notably Le Corbusier's chapels at La Tourette (1960) and the Church of Saint Pierre at Firminy[2] (2006), which combine colour, form and light. In the sacristy at La Tourette, three sloping oval rooflights are painted in red, blue and yellow, and modify the light penetrating the chapel. Similarly, at Firminy, the soffits of slot windows and high-level punctured rooflights are intensely coloured with green, red, yellow and blue, set against the untreated concrete surfaces. Holl's interior is painted white, tempered by the use of roof-lit funnels, which gently tinge the internal surfaces as the external lighting conditions change. The perception of the space is transformed as light bounces against the curved surfaces, capturing the different qualities of light from the four cardinal points by a series of angled roof lights. The volumes are shaped to further modify the internal experiences as one moves through the church, and simultaneously correspond to specific aspects of Jesuit worship.

Holl's analogy is of bottles acting as containers, in which light is captured and stored, with each volume differentiated in form and reflected colour. In his early conceptual watercolour, the vessels are set in a box, articulating the various stages of the worship within the unity and security of the church. Large study models were made to try to replicate the effect. Colour and light can be highly deceptive on models; one cannot easily scale light in the same way as spatial dimensions, so the final effect is never entirely predictable.

Watercolour conceptual study for Chapel of St Ignatius, Seattle (completed 1997), Steven Holl

Each bottle is ascribed a colour, activated by concealed painted panels under changing conditions of natural and artificial light. The indirect reflected colour is coupled with a direct source in the spectrally opposing colour, generated by light passing through a tinted lens of glass. Red is paired with green, a yellow field with blue lens to the east and a blue field with a yellow lens to the west. The complex geometry happens above eye level with the volumes sub-dividing the unified rectangular spaces of the church. The colour is subtle and moves across the surfaces during the day and with the seasonal variation in the altitude of the sun.

Mixing light is, of course, entirely different to mixing paint. Researchers at the Institute of Colour and Light at the Zurich University of the Arts in Switzerland (ZHdK) have developed a range of teaching aids and installations to demonstrate the effect of the superimposition of light sources, and of contextual light conditions, on perception.[3] One consideration is the distortion of perceived pigment colour under different light sources. In an installation called *White to White*, a

Installation/teaching device *Pigment Carpet*, at the Institute of Colour and Light at Zurich University of the Arts, Switzerland (ZHdK) (2010–11), to demonstrate the additive effects of coloured light sources on pigment. The floor is painted with two adjacent panels of white and red paint. White light is then projected onto the red pigment and red light onto the white pigment, with the outer corners left without light as a control area for calibration

Base projection superimposed with:
(top) blue light
(middle) green light
(bottom) cyan light

series of vertical panels painted in Natural Color System (NCS), off-white tones (greenish white, bluish white, and so on) can be observed to shift subtly in appearance as the light source is varied from warm white to cool white light.[4] A more dramatic installation, *Pigment Carpet*, demonstrates the effect of additive light superimposed on a floor surface, with adjacent panels of red and white pigment. A data projector then sweeps the surface with coloured light, generating pattern and vivid change as the same light source is viewed against the different pigments and as coloured light sources overlap. These installations are simultaneously poetic and highly instructives and are considered part of the dissemination of research into the inter-relationships between colour and light. Josef Albers considered the additive properties of coloured light to be firmly in the realm of the physicist and not the direct concern of the colourist, but made a distinction between 'factual fact' from 'actual fact' in relation to our perception of colour.[5] 'Factual' is based in measurable physical reality; for example, the wavelength of red light. 'Actual' is that which is perceived. His use of 'actual' indicates the strength of the illusion and proposes that cognitive deceptions are universally acknowledged and easily repeatable, provided that the viewer has normal colour vision.

Temporality

STEVEN HOLL'S use of coloured light as a temporal phenomenon enacted by sunlight, daylight or artificial light sources is a fundamental interpretation of colour as light, reflected and refracted through space and varied by time. Alberto Gomez-Perez, in his introductory essay to Holl's book *Intertwining*, suggests a connection with Buddhism in 'the unreliability of the present instant which is continuously transformed into past and non-being'. Holl's preference for indirect, tinted shadows inflecting otherwise white surfaces purely by reflection and dissolution is appropriate to his concern with the transient experiences in space and the metaphysics of light:

> The merging of object and field yields an unmeshed experience, an interaction that is particular to architecture. (e.g. Colours of stone revealed by direct sunlight, gone in shadow.) Unlike painting, we can turn away from, and music and film we can turn off, architecture surrounds us. We must consider space, light, color, geometry, detail and material in an intertwining continuum.[6]

The Cranbrook Institute for Science (1992–99), designed by Steven Holl Architects as an extension to a celebrated Eliel Saarinen building (1942) in Bloomfield Hills, Michigan, USA, is linked to the existing building through an axis that Holl refers to as the 'Stairway of Inexplicables'. The building is centred on a garden where scientific phenomena are explored through a series of pavilions such as the 'House of Ice' and the 'House of Vapor'. Not only is the physical and bodily manifestation of these experiments fully explored, the very use of language to label spaces by association with particular phenomena is indicative of his desire to engage the senses haptically and to raise awareness of the connection between natural sciences and architectural space.

'Light Laboratory' entrance foyer (top) and red walls in circulation areas (left) at the Cranbrook Institute for Science, Steven Holl Architects (1999)

The entrance lobby, painted solely in white, is one of this series of experiential and immersive spaces, in this case a 'Light Laboratory' that has a spectacular south-facing facade subdivided into irregular panels, each with a different type of glass.[7] The varied optical qualities of the glass distort, focus, and refract the changing light conditions, producing moments of intensity, shadow, streaks of light, blurring, diffused patterns and the occasional spectral projection through prisms built into the facade. These work in concert to animate the space as part of the highly experiential series of internal and external spaces. This is a subtle colour palette; the same base wall is modified in appearance solely by the instrument of light and by the passing of time.

Elsewhere in the building, Holl uses deep red ochre on plywood surfaces set against black concrete floors.[8] This is unusual for this architect, who tends to prefer white walls and simple surfaces. A key principle of Steven Holl's architecture is the 'spatial energy' generated by the interaction of the body moving through space.[9] His use of colour and light allows each work to be open to interpretation by the mood and perception of the user. The experience is therefore not entirely predictable or under the control of the architect, as would be more the case if using pigmented surfaces. Colours applied with pigment, while still entirely dependent on light conditions, context and material surface are, by comparison with coloured light, essentially more stable.

One of Holl's best-known small buildings is an office building located in the Sarphatistraat, Amsterdam (2000). Again, Holl uses reflected light from a pigmented surface or filter, which is deliberately concealed and activated by the play of light. The colour attracts the eye and obliquely reveals a hidden space beyond that which is immediately experienced, adding a layer of ambiguity. Holl further celebrates this element of doubt by using layers of different materials, some perforated, some solid, some translucent and some clear. The layers seldom entirely align, and are sandwiched in different combinations to distort the surfaces and indicate depth. The hidden, coloured surfaces produce reflected panels of colour that appear to hang in space. One can see connections with Gerrit Reitveld's use of colour as a constructive element, disembodied from the unity of the structural space. But here it is even more ethereal because one cannot see the panels themselves, only their reflection onto other surfaces, or through the veil of the porous surface. The effect is intensified at night, when artificial lighting projects the coloured rectangles onto the surface of the adjacent canal. The colour is so strong against the dark water that it appears to become fused with the liquid. Colour and light are therefore employed as part of an experiential, multi-sensory architecture, the architect deliberately abdicating a degree of authority to natural phenomena.

Holl has recently completed a small building in Norway, which took over 15 years to reach fruition. The Knut Hamsun Museum (2009) celebrates the work of the Norwegian author, interpreting themes drawn from his literature in architectural space. Holl is known for a tendency to use oppositions in his architecture. Counterparts such as light and dark, white and black, day and night, open and closed geometries, Apollonian and Dionysian archetypes of order and disorder, rationality and emotion, subjectivity and objectivity. The Knut Hamsun project contrasts a black stained timber exterior, reminiscent of the ancient timber churches of the region, with a pure, white-painted in-situ concrete interior. Daylight activates the spaces. In this very northern latitude, the extremes of sunlight vary from a period lasting from December to February with

Knut Hamsun Museum, Norway,
Steven Holl Architects (2009)

no sunlight, to the summer months with no darkness. Holl is clearly invigorated by designing in such extreme dualities of light and dark. He notes the extraordinary horizontal light in summer, which penetrates the interior through carefully placed apertures when the sun skims along the horizon.

Canadian artist/designer Doreen Balabanoff has observed that the movement of sunlight across architectural space often goes largely unnoticed when the light is white, but our awareness suddenly heightens when the light is coloured. There are, of course, many notable exceptions to this, as architects are very aware of the significance and power of natural light. Balabanoff suggests that the more time we spend in urban conditions, indoors, under tempered light conditions, the less we are aware of natural cycles of seasonal and daily patterns of light.[10] Sadly, some are simply ignorant of the seasonal sunpath, altitude and variation, yet such simple rhythms are known to be linked to our sense of well-being.[11] Balabanoff's *Colour Chords* installation in Dupont, Toronto, Canada (1992), made use of the transient qualities of moving coloured light to reconnect the building user to daily and annual sun cycles and to light's interplay with space, place, form and time. As an installation, it is dramatic and immediate, but also requires the observer to wait, to be patient. The coloured shafts of light cross the wall, floor, columns and beams, and invigorate the space. Subtle ambient effects of spreading colour, material reflectance and coloured shadows further enhance

Winter solstice, 11.30am (left), and summer solstice, 11.30am (right),
Colour Chords installation in Dupont, Toronto, Canada,
Doreen Balabanoff (1992)

space/light awareness. Balabanoff is intrigued by the elemental qualities of colour and light and has situated related works in various settings, including hospital and public school buildings. These projects are entirely site-specific, honed out of observation of an existing situation, activated by simple insertions of coloured, mouthblown glass as filters. The temporal qualities of these shifting colours both engage and calm the occupants.

A further consideration in relation to daylight is that variations in light conditions have a direct effect on the perception of colour pigments. When light grows dim, colours fade and forms flatten. We may still 'see' colour, but the colour may be purely a memory or an association with an object, a phenomenon known as chromatic adaption. Leonhard Oberascher, an Austrian artist and psychologist, has systematically observed and recorded the dynamic shift in colour pigment caused by sunlight in outdoor conditions. An experiment involving taking photographs of a simple cellular box, painted on the interior surfaces, every five minutes over an entire day and cataloguing the images in relation to data on the exact time, azimuth and altitude of the sun, makes the shift in the observed colour relative to sun position extremely evident. The project is a clear demonstration of Albers' 'factual fact'–'actual fact' and the elasticity of colour. Daylight, coupled with open surroundings, is also known to make exterior applied colour appear lighter than the same colour used internally. Internally, corners of rooms intensify colour as the reflected light waves from adjacent walls produce interference, just as water waves overlap and reflect from obstacles. Oberascher argues that cyclical alterations in the perceived colour are rarely used intentionally in architectural design, yet are evidently present as part of our physical and psychological sensing of time and place. The extraordinary elasticity of colour under different light conditions adds a further dimension to site-specific architecture. Not only can colour define place, but it can define multiple readings of the same place depending on the time of day and the season.

(top) Elasticity of colour demonstrated in Leonhard Oberascher's experiment on the *Interaction of Colour and Light* (2010);
(bottom) detail, in which the successive images clearly indicate the change in appearance of one particular hue

Location and translocation

STUDIES by the colour theorist Jean-Philippe Lenclos, who pioneered the documentation of colour on the exterior of buildings in the 1980s, record regional variations that support Kenneth Frampton's concept of 'Critical Regionalism'.[12] Lenclos catalogues the pigments in both applied and natural materials by region, starting in his native France then extended to curate a *Geography of Color*.[13] His work has served to document the palettes of particular localities. The method is used extensively by city planners in sensitive settings, for example in the historic centre of Turin, Italy, to control the choice of hues for render, shopfronts and adverts. The implication of this approach is that colour, even applied colour, should be responsive to its physical context and cultural traditions.

Building portrait of housing at Diggelmannstrasse, Zurich by Gigon/Guyer Architects as one of a series of documentary studies by the Haus der Farbe, Zurich, in *Farbraum Stadt: Box ZRH*, 2010

In the case of Lenclos, the argument is that historical architectural colour was most likely to be driven by locally-sourced, natural pigments, mostly mineral, but is also a significant marker of place and should be respected. In historical settings that have a tradition of exterior applied colour, such as the west coast of the USA, central Europe or Scandinavia, colour may be regulated to preserve cultural difference, as well as the memory of a specific place. The artist David Batchelor notes a micro-regional difference between parts of London: 'There is no interesting colour in Knightsbridge – for colour in the city, go to Hackney.'[14]

The Haus der Farbe, in Zurich, has undertaken an extraordinary, detailed research project, cataloguing colour used in the built fabric of the city. Conceived in 2004, and piloted in one district in 2007, 41,000 buildings have been surveyed. Working with architecture students, over 800 colours were documented and then reduced to a colour fan with 115 shades in 12 colour families matched to Natural Colour System (NCS) colour codes. The conclusions have been published in the form of a book of essays and a series of abstract postcards – *Farbraum Stadt: Box ZRH*.[15] Each building 'portrait' was hand-painted with the area of each colour defined in proportion to that observed on the building. They were aware from the outset that dynamic factors such as the time of day, the weather, the season and the individual judgement of each student would temper the record of each building and, to some extent, subvert the apparent precision of the method. The colours documented include painted stucco and contemporary coloured glass, as well as embedded pigments in roof tiles and other building materials. On the reverse of each card, the building is catalogued. This is a comprehensive venture, minutely analysed, and aims to uncover historical trends, differences in districts, and predominant shades. Their intention is that this exercise will not be used to control or legislate colour use, but to provide a platform for informed discussion.

Offices at Monterey, Mexico,
Legorreta + Legorreta (1995)

Time will tell but, encouragingly, the study includes the highly colourful palettes of recent projects such as those by Gigon/Guyer and by Mueller and Sigrist, as well as traditional buildings.

Flying over Madrid, it is extremely clear that the colour of the city matches the pinky sand colour of the surrounding landscape, which is dotted with dark green trees. The British architect David Chipperfield, when designing housing at Villaverde in Madrid (2005), adopted the same palette to tint the precast concrete panels of his social housing project five shades of terracotta pink. To the architects, the panels were an expression of giant bricks, soft milky colours that appear paler in the intense sunlight. The massive Barcelona City of Justice (2002–2009), by the same architects, also uses tinted concrete, but in various less-natural shades. These projects are unusual for the office, which is known for a restrained, natural palette and white-painted interiors. Chipperfield's modernist approach is exemplified by the renovation of the Altes Museum in Berlin (2009) and the nearby Am Kupfergraben 10 gallery (2007). Was there something about the intensity of colour and light in Spain which prompted this shift in the architecture, or was it an intuitive emotional response to the site?

The South American architect Luis Barragán's intense pinks and oranges are chosen only after lengthy observation under the weather and light conditions of the specific site. In his case, the context of climate plays a part in the selection, although he admits that the source of the hues may be arbitrary, some taken from paintings.[16] The colours for a purple and pink house originate from

Haus über Graz, Austria,
Szyszkowitz + Kowalski Architekten (1974)

Student centre at the Technischen Universität, Graz, Szyszkowitz + Kowalski Architekten (2000)

a painting by the Mexican artist Jesús Reyes Ferreira (1880–1977), known as *Chucho Reyes*. Barragán's colours are therefore both dissociated by source and associated by specific context of culture and light. The South American contemporary architects Ricardo Legorreta and Victor Legorreta refer to the same artist as a source of the intense colours inherent in their architecture, noting that the artist's 'use of colour was irresponsible' – meaning positively so. Ricardo Legorreta, based in Mexico, is known for the use of highly saturated applied colour on stucco facades. Deep reds, yellows and purples are driven by cultural tradition but also by the intense quality of the light.

Central European architecture has a tradition of painted stucco facades, so it is perhaps unsurprising that architects Szyszkowitz + Kowalski see nothing particularly meaningful in their use of colour externally, yet it is a significant element in the work of the practice over a sustained 40-year period. Early projects, such as the Haus über Graz, Austria (1974), which is clothed in a vibrant purply-pink coated steel sheet, are not remotely deferential to the historical regional architecture. Yet, the way in which their buildings modify and nestle into the site, using the natural topography but also carving new external spaces by modifying the land, results is a very contextual and, some might argue, regional architecture.[17] Colour seems to come so naturally to them that they are reluctant to acknowledge its powerful associations. A more recent project for a student centre at the Technischen Universität (Technical University), also in Graz, uses a turquoise green on the urban perimeter faces, changing to a startling orangey-brown within the courtyard. Szyszkowitz + Kowalski begin every project with hand drawings, as does Steven Holl, teasing out the essence of each site and place. These highly colourful studies have perplexed architectural critics, struggling to pigeon-hole their unique, expressive and socially-driven architecture into a particular genre.

Frank Werner's use of the term 'translocation' to describe the work of Szyszkowitz + Kowalski serves to emphasize their particular ability to create places that are cultural and socially significant, and highly specific to the site. The creation of such places is a combination of the manipulation of landscape, planting and sculptural architectural design, but also makes use of colour to modify, emphasize or embed in the site context. In many ways this is similar to the approach of Steven Holl, although the architecture is very different. Both consistently approach their architecture from an artistic starting point.[18]

In addition to his built work, Steven Holl is a prolific writer. His written output identifies a number of themes in his work, which correspond with the development of key projects, or series of projects. The first book, *Anchoring*, is concerned with the vital contribution of site and place. There is, however, little or no sense of critical regionalism in the way in which he embeds the building within the site. It is a physical and natural rooting, a topographical connection with the ground, a visual connection with daylight and a metaphysical foundation to each work. Holl's attitude to site has been compared to that of the Portuguese architect Alvaro Siza.[19] Neither architect is afraid to alter the site substantially in order to situate the building more effectively. Holl's use of light, either white or tinted, is clearly informed by this experiential grounding. His use of pigments, however, is more likely to be driven by a cultural context or an abstract conceptual model, and is less easily explained.

Precision or doubt?

STEVEN HOLL has recently noted that ideological frameworks should not be carried directly from one project to another, firmly believing that each project must respond to its unique brief and to the social, cultural and economic context in which it is sited.[20] His projects are, however, linked through a series of principles. In the Sarphatistraat Offices in Amsterdam, Holl explores concepts of porosity utilizing copper mesh, folded to imply solidity and volume, while in reality it is a thin porous surface. The project also plays with less tangible considerations. Holl has a fascination

Offices on Sarphatistraat, Amsterdam, Steven Holl Architects (2000)

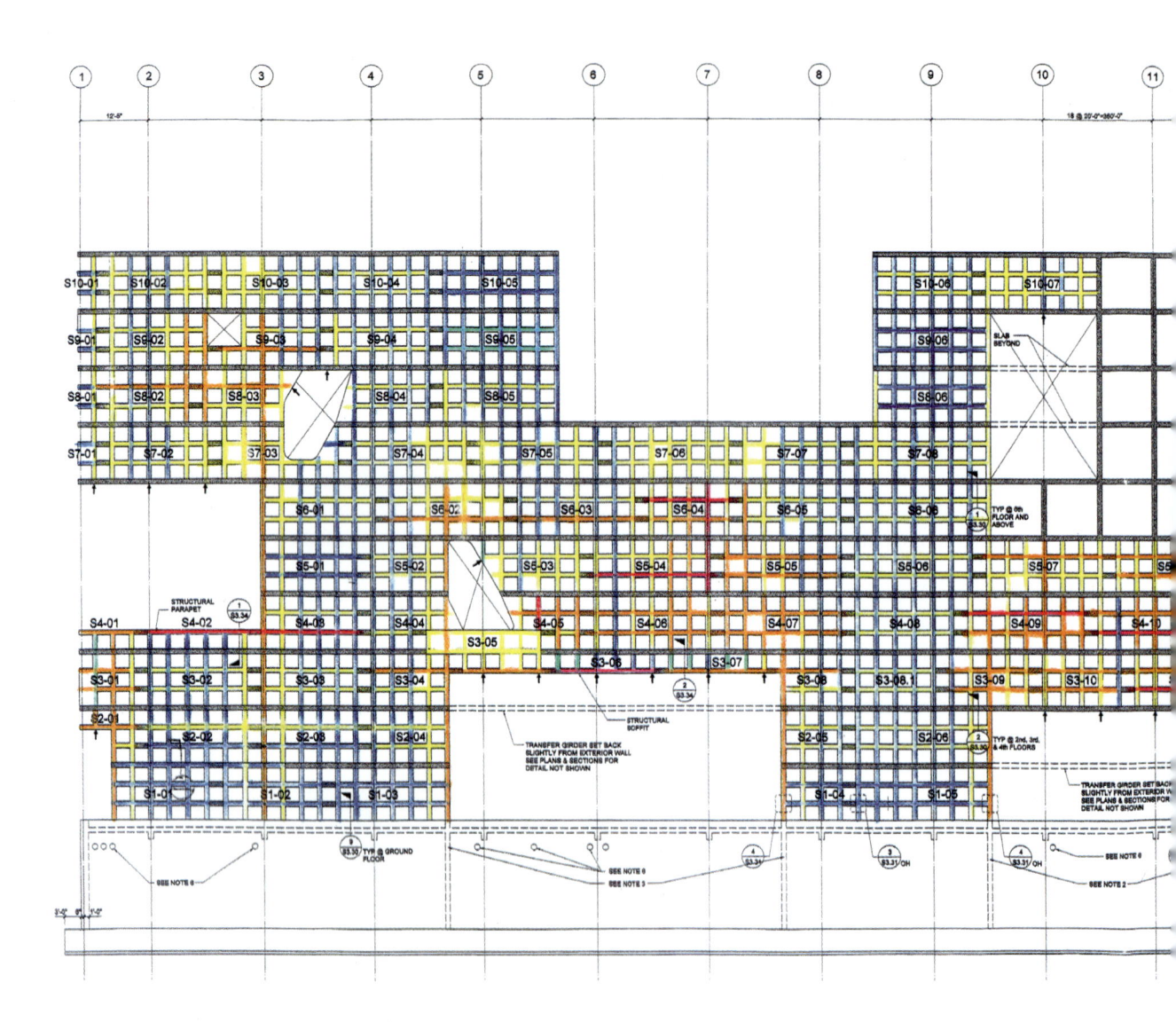

Analytical study drawing by engineers Guy Nordenson of the maximum stresses in the south facade structure of the Simmons Hall of Residence at MIT, Cambridge, MA, USA, Steven Holl Architects (2002)

FRAMING ELEVATION
AT SOUTH WALL - LINE A

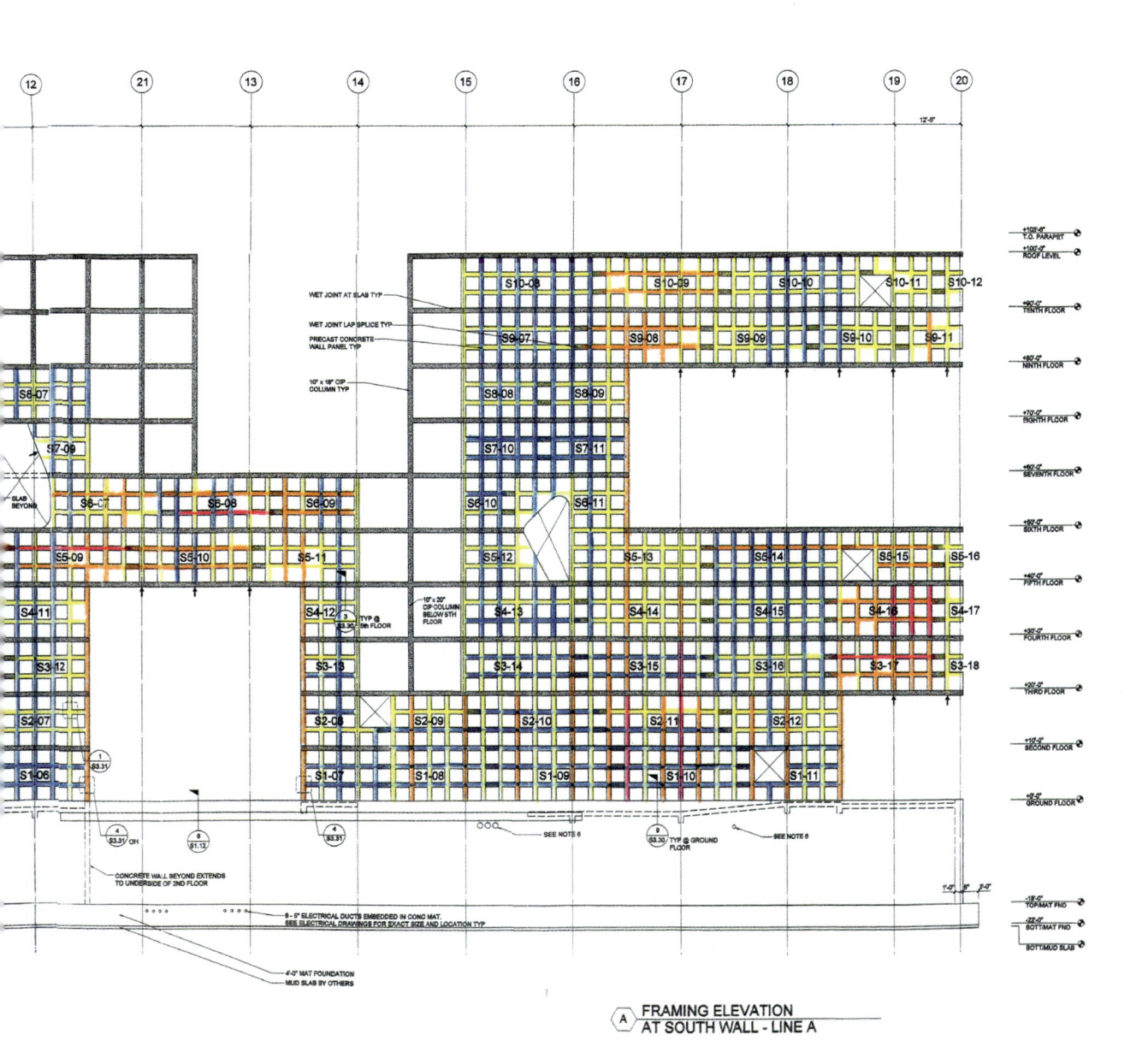

with mathematics, proportion and the golden section. To the physicist and mathematician, colour can be expressed in precise numerical equations, according to the wavelengths of each hue. The visible spectrum is, of course, only a small proportion of the full array of radiant waves. Holl's interest in metaphysics extends to mathematically generated rhythm, proportion and fractal geometry. His projects couple regulating grids with amorphous sinewy voids. The use of grids is evident at the Simmons Hall of Residence at MIT, Cambridge MA, USA (2002), the Sarphatistraat project in Amsterdam and the Linked Hybrid apartment buildings (2009) in Bejing, China. The Amsterdam project was derived loosely from a Mengel Sponge, a fractal, infinite cube in which holes are formed making the grid porous. If carried to its pure mathematical form, the porosity is considered to terminate in a form that has zero volume. Holl also makes reference to the music of Morton Feldman, most notably his 2004 composition *Patterns in a Chromatic Field*, in which fugal repetitions scale and morph. As with many generative concepts, Holl's abstract sources may be lost on the observer, but the resultant proportion, scale and rhythm are controlled.

In the early 1960s, a mathematical toy, Cuisinaire Rods, was a popular kinesthetic tool to help children learn maths. Each number from 1 to 10 was ascribed to a small block of wood, proportional to the value of the number but also stained with bright dyes. The number 1 was a small white cube, 2 (being a key prime number) an intense red, 3 a lime green, 4 pink, 5 yellow and so on, until reaching 10, which was orange. Combining a red and green rod end to end, aligned in length with a yellow 5. The mathematical relationships were visually apparent and lodged very firmly in the mind through the senses of touch and sight. A secondary use, but one that was very satisfying, was as a construction tool with each component perfectly proportioned by geometry.

For Holl, mathematical relationships establish a relational scale, which he considers satisfies a spiritual longing in architecture, which is beyond the physical. Le Corbusier argued similarly in developing the 'Modulor'. The Simmons Hall of Residence at MIT provides accommodation for 350 students; the rhythmical register seems contrary to the human scale of the individual room, however, and is deceptive. Holl has been accused of an overly willful imposition of the porous grid in the dormitory that is not wholly successful in making accommodating, habitable space. The applied colour also seems somewhat arbitrary. The explanation for the colour given by Holl is abrupt and ambiguous:

> Based upon a structural diagram used to co-ordinate the size of reinforcing steel in the Perfcon panels, the colored jambs express the anticipated maximum stresses in the structure. The colors reveal the size of the reinforcing steel cast within the Perfcon panels. Blue = No. 5, Green = No. 6, Yellow = No. 7, Orange = No. 8, Red = No. 9 and No. 10. Uncolored areas are No. 5 or smaller.[21]

As the colour is applied to window soffits and jambs, from certain viewpoints it is invisible. Holl does not explain why these particular colours are chosen for the codification. It seems that there is either more at play than is explained, or that Holl himself is uncertain. This was Holl's first large-scale use of applied colour. Without the clear yellow, red, blue and green, the block is unrelentingly grey and austere externally. Colour also emphasizes the surface of the largest voids and undermines the regularity of the grid, externally, in a playful way. Similarly, amorphous forms are used internally to penetrate vertically through floors and subvert the grid. The colour used in his Linked Hybrid in China is somewhat unconvincingly argued as being derived from polychromatic Buddhist temples, then applied using the *I Ching*. As an attempt at rooting a project in a culture, it seems superficial. Holl, it would seem, is caught between an obligation to offer a rationalization for his use of colour, while simultaneously proclaiming his joy in ambiguity, and doubt.

Space, place and time

Steven Holl's architecture is characterized by a driving ambition to create architecture in which the user is engaged as a participant, experiencing space at a deep and lasting psychological level. The unreliable, ephemeral and unpredictable nature of colour is an integral part of his phenomenological approach. He is prolific both in writing and in built work, with each project contributing as an heuristic device through which he experiments and tests relationships between space, place, time and light. Although colour is often present, it is rarely a focus in itself, but seen as part of the immersive capabilities of architectural space to trigger deeply felt, emotional responses.

For Holl, one of the most significant regional influences is the quality of natural light. If one were to transport Legorreta + Legorreta's intense hues to Norway and Sweden, they would be staggeringly overpowering in the gentle natural light. When Grete Smedal was commissioned by the Spitzbergen Coal Company to design a colour codification and palette for the settlement at Longyearbyen, Norway, at a latitude of 78° North, she developed a palette of colours based on the Natural Color System (NCS) triangle, which have almost equal values in terms of whiteness, blackness and the intensity of the chromaticness, and applied these to the external surfaces of the individual buildings.[22] The new colours, while not specifically sourced locally, were chosen to respond to the quality of light. Smedal's imported colours now define this place.

The four-dimensional meaning and experience of place includes the colour of the specific surroundings, seen in relation to the passing of time. The authors of the Zurich colour study asked people which colours they associate with the city. Although the analysis of the buildings clearly indicated a predominance of beige, greys and ochres, the citizens thought of their town as green and blue. It seems that the natural features of trees, river and lake are highly influential in establishing the visual character of a place. One can see this as highly significant in relation to perception. The immaterial, imagined city challenges physical reality. The art historian John Gage has documented the cultural significance of colour, symbolically and by association. Surface-applied colour has an immediate effect on the reading of form, but will also have an effect by imbuing the context with a sensory tuning that gives character and identity to the location. It is not merely the immediate emotional response to a place that can be affected by colour, the cognitive memory of the place is likely to register colour spatially and by association. As noted by Faber Birren, 'technical considerations are less important than the evidence of vision'.[23]

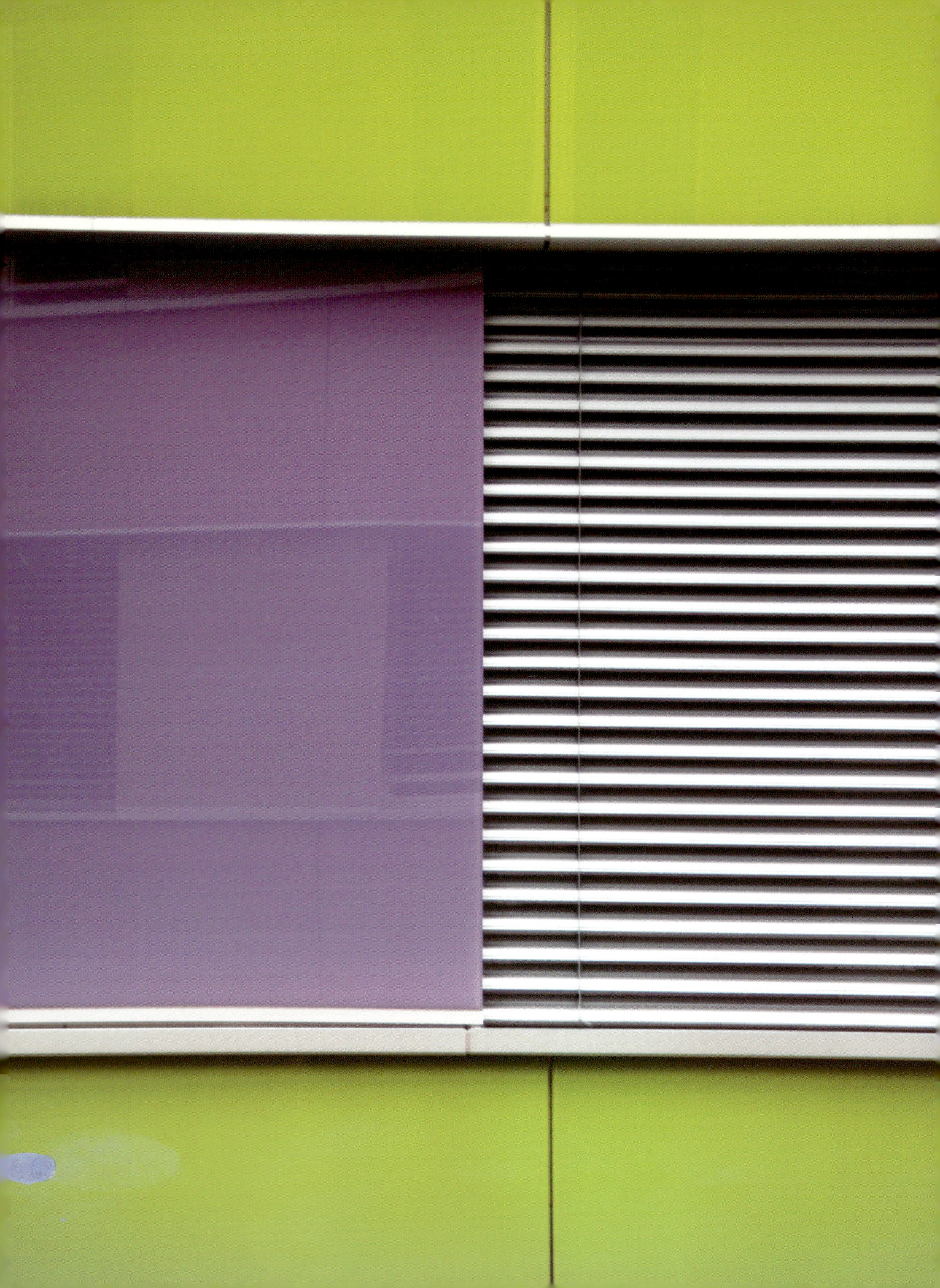

– 7 –

Surface and edge: Gigon/Guyer

THE architecture of Swiss-based architects Annette Gigon and Mike Guyer is visually arresting. Colours, where employed, are often strong, vivid and unsettling in their context. A number of their recent buildings use highly polished coloured glass, making surfaces that reflect their surroundings and give little insight into their contents. Others employ pigments either mixed deep within the construction, or laid in thin coats clearly applied to the surfaces and exposed at the edges. The buildings, being predominately private, are experienced first and foremost through their visual appearance rather than spatially. Juhani Pallasmaa, in his book *The Eyes of the Skin*, argues that architecture should stimulate all senses, not purely the visual, and bemoans the gradual hegemony of the ocular sense in contemporary art and architecture. This preference for a phenomenological approach is common to other theorists, such as Christian Norberg-Schultz, who suggest that architecture should engage the body to allow the experience of space and place to be sensed holistically. It is interesting, therefore, that Gigon/Guyer should adopt a phrase most often associated with Norberg-Schultz – the *'genius loci'* [1] – to describe the importance of place in their work.

Genius loci

IT is not immediately obvious how Gigon/Guyer's use of intense colours and reflective materials relates to the reading of the site. To understand their meaning of place, one must therefore look beyond the purely visual. As Charles Rattray and Graeme Hutton have observed, 'One of their [Gigon/Guyer's] guiding principles is that place and type are remembered formally and materially, then selectively reconstructed'.[2] As an example, Rattray and Hutton suggest that the 'Profilit' glazing of the Winterthur Museum can be read as part of an industrial heritage, and the earth-brown walls of the housing at Kilchberg as expressive of an agricultural tradition contributing to 'a reservoir of visual catalysts', which 'if not uniquely Swiss – might be thought of as place-specific.'[3]

facing page
Facade, Diggelmannstrasse, Zurich, Gigon/Guyer (2007)

Gigon/Guyer seem to embark on a forensic interrogation of each site, searching for elements that could be seen as part of the essence of a place, but which they then re-present in a form that, particularly to the untrained eye, is not evidently of its site. As with a developing infant, their designs may start out with a genetic framework, which gradually absorbs information from its surroundings. As the design develops, however, a different character emerges. It may bear some hereditary mannerisms from its origins but may also supplant the site to the extent that the new building becomes the most significant memory of the place. It has been noted that their Sports Centre in Davos (1996), with its supergraphics and strong underlay of orange, blue and yellow, has become emblematic of the place. Here, the interior of the building is polychromatic with six colours: raspberry, turquoise, lime green, dark blue, white and apricot. Externally, strong orange, blue and yellow panels lurk behind a layer of untreated slatted timber. The intense colours of the Sports Centre contrast with Gigon/Guyer's monochromatic Kirchner Museum (1992) in the same town. Their reading of place, therefore, is inseparable from the function of the building.

The most poetic example of this rhetorical process is found at the Villa am Römerholz, Winterthur (1998). Precast concrete panels, used as cladding for their extension to the Oskar Reinhart collection gallery, are created with ingredients that include powdered copper and Jura limestone – being the predominant materials of the existing buildings on the site. In the extension, the copper is powdered and used as a pigment in the concrete mix. Making good concrete has been compared to baking. While cooking can be imprecise to some extent, baking needs greater precision and understanding. The ingredients can be varied, but the recipe is essentially similar and the 'flavour', appearance, durability and colour are therefore utterly dependent on the mix. Adrian Forty has argued that it is a mistake to designate concrete as a material: 'Concrete, let us be clear, is not a material, it is a process'.[4] Its indeterminacy is its very nature. The limitations and possibilities of the process are dependent on interpretation by architect and contractor. At Römerholz, the choice of precast panels was dictated by the need to experiment with the mix before finalizing the proportions. The roof of the new extension is also copper and is detailed to invite rainwater to drip down the face of the exterior walls to assist in the oxidation of the copper in the concrete. The process is described by Gigon/Guyer as making 'a kind of journey through time to both of the older historicizing building elements, in the sense of an "alchemistic" adaptation of the new building to the genius loci'.[5]

In Zurich, Gigon/Guyer have adopted a similar experimental approach for a small 'Stellwerk' (railway switching station) set immediately beside the railway tracks, in what is a fairly hostile industrial environment. The colour of the concrete, this time cast in-situ, is derived from a conceptual analysis of place. They noticed how the rust from train wheels and tracks pollutes concrete sleepers, sidings and even plants along the line, and is particularly acute where trains decelerate. The rusty colour inspired the use of iron oxide pigment mixed into the concrete. The reason, as observed by Martin Steinmann in his essay 'Conjectures on the architecture of Gigon/Guyer', is 'not to camouflage a building of this size; I think the reason was to insert the building into the experience that we carry from this world, and to make us conscious – again – of this experience'.[6] The project was a collaboration with artist Harald F. Müller, who also sourced a strong red, blue, yellow and dark brown, used internally on built-in furniture, from objects found in the immediate

Pigmented concrete at the Stellwerk, Zurich; the concrete has gradually changed from a deep red-brown to a pinky-orange, Gigon/Guyer (1999)

surroundings of the site. Lengthy trials were undertaken to achieve the 'correct' colour for the concrete walls, before finally opting for a dark brown pigment in a 9 per cent admixture. Ironically, time has affected the colour, which is now a much more vivid orange. Gigon seems perfectly comfortable with the metamorphosis, despite the search for initial accuracy in the tone. It is a reminder that colour is unstable, transient and can be unpredictable. Both projects are evidence that Gigon/Guyer have an interest in experimentation with materials. Not only that, but the belief that the collective memory of a place should be invoked through an appropriation of its material heritage.

The analogy with cooking, or indeed, baking, is picked up in a published interview with Annette Gigon in *The Architect, the Cook and Good Taste*.[7] The relationship of cooking to a sense of place, or region, is discussed and Gigon reiterates the importance of location in their work as one of the ingredients. She cites as an example of form rather than colour the Museum Liner in Appenzell (1998) where the saw-tooth roofs were derived partly from its mountainous outlook. In terms of constraints to design, the local regulations and legislation, Gigon notes, can be seen as the most decisive 'measures' in the design mix.

Both the Stellwerk, albeit unintentionally, and Römerholz, quite deliberately, accept ageing as a natural process in architecture. In France, at Mouans-Sartoux, an ordinary place is made extraordinary by the vivid lime green of Gigon/Guyer's gallery building Donation Albers-Honegger (or EAC, 2004). Its shape and colour are said to be derived from its location, set within a cluster of trees. The intensity and brightness of the colour is startlingly alien yet, at the same time, analogous to the hues of the natural site context. Aware that, over time, algae would cover the building, the strong green was chosen to ameliorate the effect of such discolouration.

Donation Albers-Honegger/EAC,
Mouans-Sartoux, France, Gigon/Guyer (2004)

In their book *On Weathering*, Leatherbarrow and Mostafavi discuss the inevitable transformation of buildings as nature gradually assimilates them into the sites from which they originated. The growth of a natural patina that modifies the colour and surface texture is a step towards eventual ruin. The idea of accepting, predicting and incorporating such change is unusual. Whether the clients will permit such deterioration, or maintain the crisp green colour, which is now iconographic of the building, is itself a matter of time.

A housing project at Broëlberg 1, in the village of Kilchberg, south of Zurich (1996) is set within rolling agricultural land and has also changed its colour over time. The original colour used on the outside-facing walls was a dark earth-brown and, as with the Stellwerk project, the colours were chosen in consultation with the artist Harald F. Müller. On the inside faces of the blocks, facing a central courtyard, a bright orange was used. The colours were seen as analogous to each other, being similar in hue, yet contrasting. The brown was intended to convey the natural earthy tones of the surrounding fields, at the same time as alluding to the (chemically) organic insulation board beneath the render. By contrast, the orange was deliberately artificial and man-made. To Gigon/Guyer 'brown is quieting, integrative, natural, restrained and noble; the orange, meanwhile is shrill, artificial, exotic and downright beautiful'.[8] Over time, however, the dark brown render proved to absorb too much heat and began to cause an unacceptable level of deterioration in the insulant and has now been repainted in a lighter brownish-grey, still maintaining the conceptual basis of the colour choice. Although the colour is applied in a painterly manner, this is a reminder that architecture is not painting, and has to accommodate just such complexities of function and durability as well as appearance.

Part of the idea of inviting staining at Römerholz came from observation of the stone podiums of bronze sculptures in the garden, stained by the rivers of water dripping from the figures above. The artist Walter Pichler made a beautifully poetic use of the rusty water gradually extending further along a concrete channel from a large, Corten door at the MAK in Vienna. As Leatherbarrow and Mostafavi note:

Housing at Broëlberg 1 in the village of Kilchberg, south of Zurich, Gigon/Guyer (1996)

> While it may be possible to see all weathering as deterioration, the production of distracting marks that dirty original surfaces, this recognition of the play of shadows and the inevitability of marking suggests an alternative interpretation, one of weathering as a process that can productively modify a building over time.[9]

Variety as identity

ANNETTE GIGON and Mike Guyer met while at the Swiss Federal Institute of Technology, from which they both graduated in 1984. Both then worked for high-profile architectural practices, known for their intelligent, critically driven approach to architecture but with very diverse output. Annette Gigon worked for Swiss architects Herzog & de Meuron, and Mike Guyer for OMA in Rotterdam. They then spent a period as sole practitioners before joining forces in 1989. Their office, based in Zurich, is set immediately adjacent to one of their own housing projects on one of the sloping hills that characterize the city. Wilfred Wang, in his essay 'The variegated minimal: remarks on the architecture of Gigon/Guyer', defines his use of this phrase to describe the work of the architects as follows:

> if orthodox minimal is about the independence of the object, absolute geometry, monochromatic surfaces, and an a-tectonic construction ... then a variegated minimal

is about complex contextual dependence, dominant geometric simplicity of form and space with minor irregularities, subtly varying poly- but seemingly mono-chromatic surfaces, mega-tectonic and tectonic construction.[10]

The portfolio of work from Gigon/Guyer is indeed varied, but with two evident strands. First, there are those projects which employ a subdued, material-based palette, such as their museum designs from the early Kirchner Museum in Davos (1992), the Liner Museum, Appenzell (1998) and the extension to the Oskar Reinhart Collection 'Am Römerholz' Museum in Winterthur (1998). All of these allow the art to take centre stage, through a series of simple white-painted spaces, while the exteriors avoid being bland, characterless containers by their poetic use of, often commonplace, materials. The second strand comprises those projects which utilize strong colours – commonly their housing projects. Gigon/Guyer consider architecture as a means of engaging with the world, and colour being one tool within the architect's toolbox that can sometimes, but not always, be employed. As Wilfred Wang notes, 'any building has a material side, it participates in the constructional and financial discourse of its time'. Wang also warns, however, of the danger of putting all contemporary Swiss architects in one basket. Herzog & de Meuron, Diener & Diener, Peter Zumthor and Gigon/Guyer, among others, have

> all been regarded as exponents of a second modernism, of a return to the minimal. The similarities are indeed striking … Yet on closer inspection, the subsumption of buildings by these architects under one position is as naïve as was the notion of a monolithic modernism.[11]

Indeed, according to Foreign Office Architects (established by Farshid Moussavi and Alejandro Zaera Polo), the *'practice of stylistic consistency'* has been rendered obsolete. What interests them is 'seeking a balance between similarity and difference', and they describe the process of progression as 'phylogenesis – whereby seeds proliferate in time across different environments, generating differentiated yet consistent organisms'.[12]

Common to the majority of Gigon/Guyer's projects is a simplicity of form derived predominately from function. The buildings are spatially clear, legible and logical. In their housing projects, as with the museums, the internal spaces are not dogmatic or prescriptive of use. Externally, through materials and their use of colour, their buildings, and the spaces generated between buildings, become ambiguous. Meanings are intentionally complex, not always clear and can therefore be open to greater interpretation.

facing page
Reflections from adjacent apartment blocks in the shiny glass surfaces at Diggelmannstrasse, Zurich, Gigon/Guyer (2007)

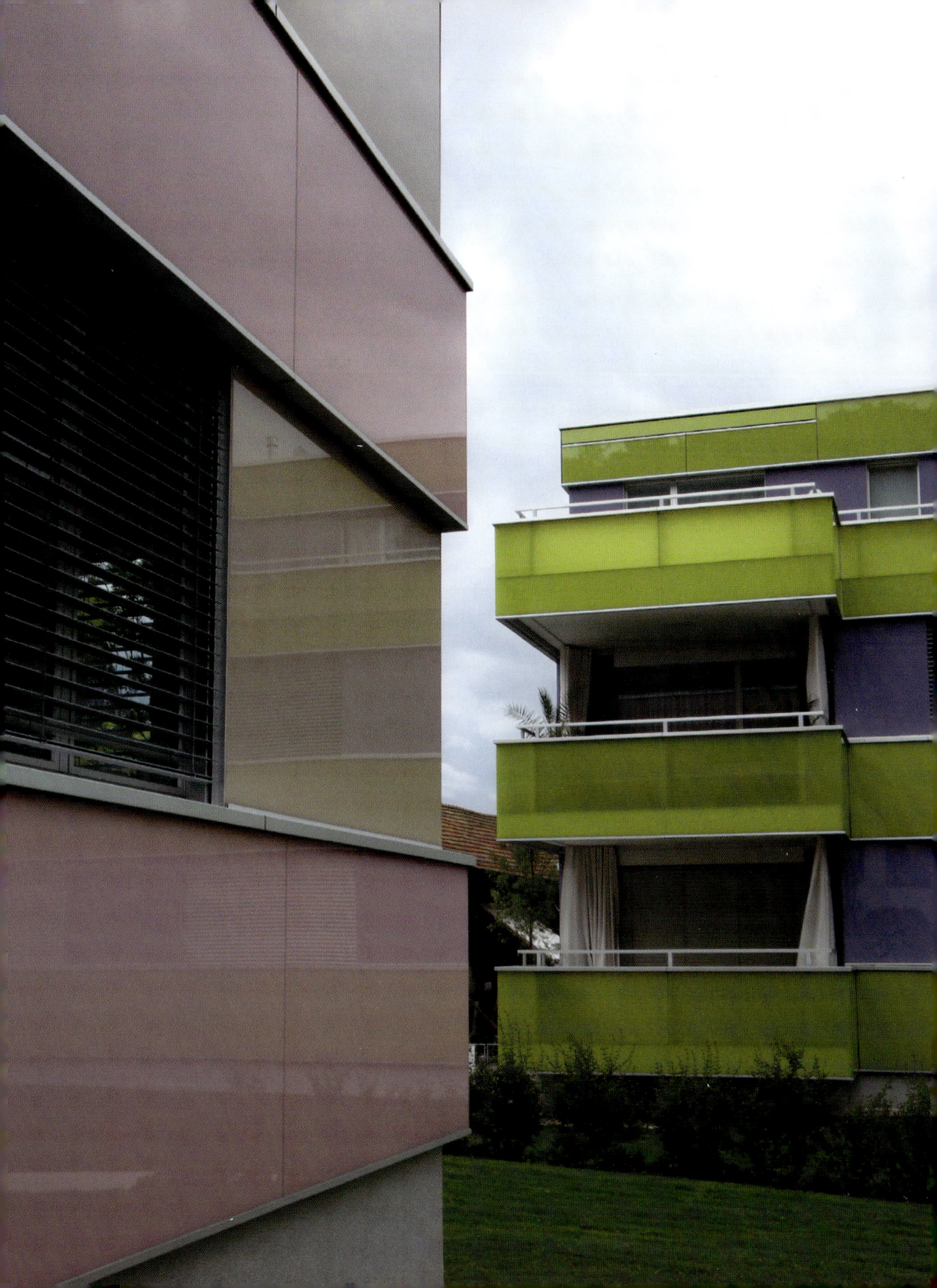

High gloss: surface reflectance

Gigon/Guyer have a strong and long-standing interest in contemporary art. A number of their commissions for galleries have been a direct result of their knowledge and appreciation of specific artists. They value the additional insight that an artist can bring to an architectural project and approached the Swiss artist Adrian Schiess after seeing an exhibition of his work. Schiess is best known for his 'flat' works where he uses highly reflective surfaces, usually lacquered aluminum, either laid on the floor of the gallery or propped against walls. The pieces are often highly site-specific, with the reflections of the surroundings, windows, people and other paintings being an essential part of the reading of his own art. Change through time is an inevitable part of the reflection, and the paintings vary constantly with the light and the people viewing them.[13] To Schiess, these panels are thought of as a part of a single huge work continually in progress. They are interactive in so far as they reflect life and invite a dialogue between the viewers and the art. Each moment they are viewed, the reflections are effected by happenstance.

The three polygonal apartment blocks at Diggelmannstrasse, Zurich (2007), undertaken as a collaboration between Schiess and Gigon/Guyer, exemplify their use of highly reflective coloured glass panels. Two blocks use yellow-green and lilac as the key colours, the other, beige-brown and pink. Although the blocks do not touch, reflections from one infect the others, as well as interacting with their immediate surroundings through reflections from beyond the site. Balconies use slightly more translucent versions of the same coloured glass, giving cloudy views of household objects and the occasional body moving in the space behind. The blocks are as alien to their setting as Schiess' reflective pools of his floor panels are in a traditional gallery. These are an extreme example where there seem to be no accommodating gestures to relate the new buildings, in material or form, to their site. The surrounding 1950s residential streets tend to adopt traditional painted render in gentle tones and a pseudo-vernacular expression. Gigon/Guyer's buildings, they acknowledge, could be seen as provocative in this location. The project 'questions the familiar environment: the colour composition is bold ... gives the optical integration a good shake'.[14]

In Brunnenhof, a large social housing project to the north of Zurich, (2007) this is explicitly not the case. The housing, for families with children, is of a scale which sets its own context in relation to parkland. There was no colour on the original competition entry but Gigon considered that the 'need (for colour) was felt very strongly'.[15] As a result, the project then became a collaboration with Adrian Schiess. A wide range of spectral hues are used in coloured glass panels, some fixed to the facades, others set in sliding rails as screening devices to the large balcony spaces which give the flats a flexible semi-external zone. The colours gradually shift and flow along each of two long blocks, following the spectrum, but with an occasional discordant panel. There is no attempt to make the blocks appear to be constructed from glass. A narrow ribbon of concrete is left uncovered to explain the structure. From a distance, viewed across the adjacent park, the

facing page
Social housing at Brunnenhof, Zurich,
Gigon/Guyer (2007)

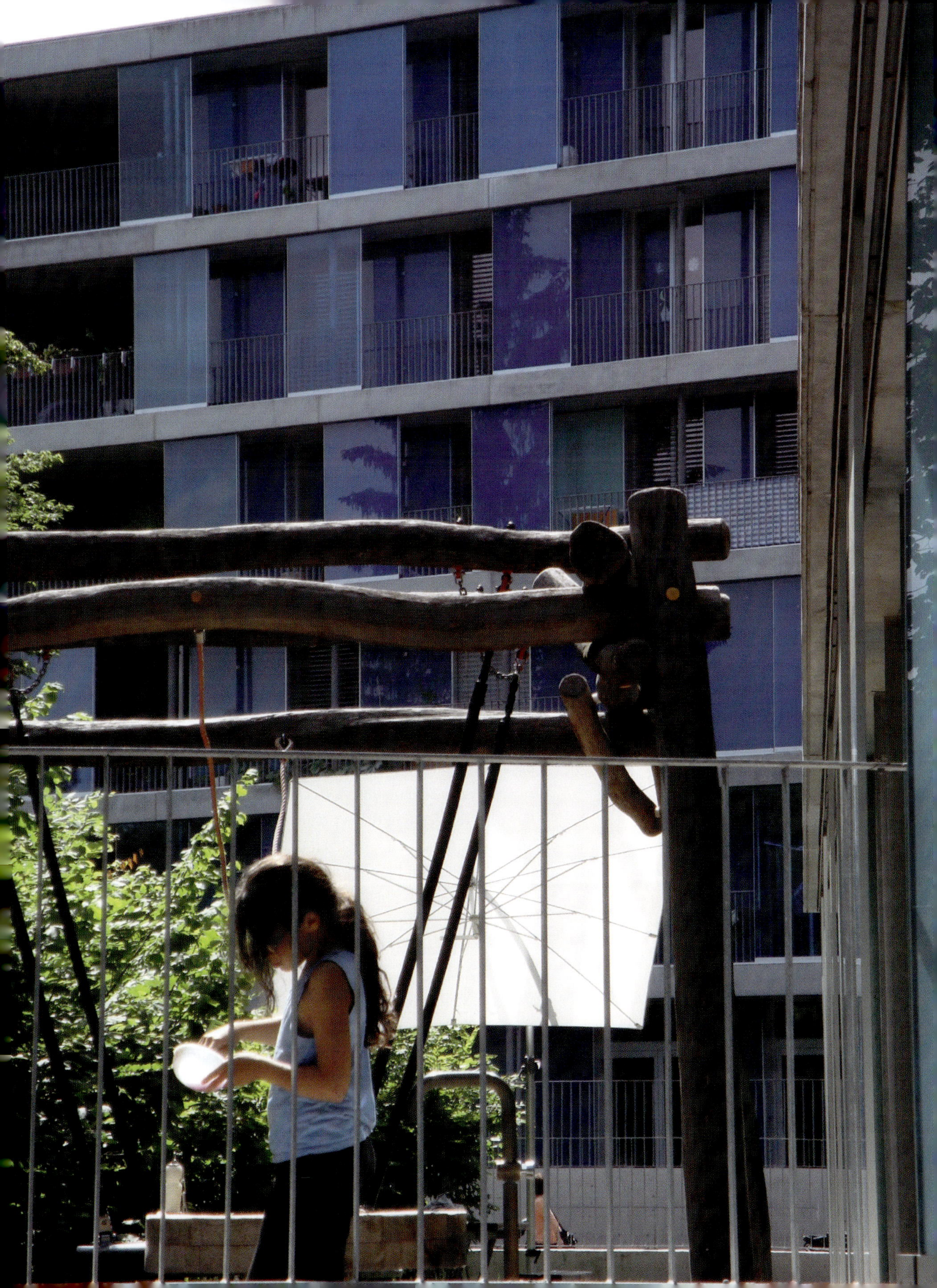

colours begin to merge. Annette Gigon has described the effect as 'a kind of expanded landscape painting when viewed from the park' and makes reference to Claude Monet's water lilies 'in which the paint turns into one ecstatic blast of colour'.[16] She reflects that they have no theories or systems to their use of colour; moreover, that Adrian Schiess is known to completely distrust any idea of system. He prefers to make intuitive judgements in relation to colour. According to Gigon, Schiess is interested in the 'dispersion' of colour, and in this project the use of reflective glass as a backdrop to the social and communal activities of the park closely resembles his own work, but on a massive scale. Each vertical panel reflects nature, life and everyday activity. Mike Guyer has suggested that glass is the perfect medium for colour, meaning that it gives colour a brilliance not possible with mineral paint. Gigon, however, is not so sure, and would argue that the reflections being dependent on weather, and changing light conditions, are less autonomous than matt powder paint. Glass is also more unforgiving in terms of the specification process being too costly to change or undertake much by way of experimentation.

Detail, social housing at Brunnenhof, Zurich, Gigon/Guyer (2007)

Of course, it is also true that reflective glass has been symbolic of a dead, uninviting corporate world, the worst offenders slipping into an implied immateriality and anonymity in a similar manner to the ability of sunglasses to conceal signs of emotion. The building reacts to its site context using two linear blocks set at an angle on plan, gently embracing the park. The front and backs are completely different in form and expression. On the sides facing the main street, a single subdued colour dominates over a thin flat facade. No attempt is made to humanize this side and it is reminiscent of an upmarket office development. It is only when the side adjacent to the park is revealed that the varying and irregularly distributed colours immediately subvert any corporate connotations with a playful, joyful nature. On this face, the depth of the balconies and haphazard signs of occupation clearly express the life and nature of the dwellings. Gigon/Guyer are aware that too much order or minimalism can be disturbed by the untidiness of domestic clutter and enjoy the process of customization, which is beginning to happen. The variation in the coloured panels is seen as being more accommodating of the changing scenery.

The most psychologically immersive of the co-authored projects with Adrian Schiess is the large subterranean lecture hall set adjacent to Karl Moser's building (1913) on the Zurich University (ETH) central campus (2002). From the main entrance, there is little advance warning of the hall below. Two pink-painted stairwells clad in medium density fibreboard extend up into the pale-green vaulted hall. The intensity of the pink increases as one descends and the daylight diminishes, to the extent that it appears to be a different colour altogether. Once inside the large room, the glossy panels and lime-green built-in benches are wonderfully lively. The effect is uplifting and intense, and students surely must feel stimulated in such an environment.

Lecture hall insertion to Karl Moser's building (1913) and contrasting pink-painted walls against the pale green original building announce the route down to the lecture hall, Zurich University (ETH) central campus, Gigon/Guyer (2002)

Externally, the extension is expressed to the rear of the building with a pinky-red tinted concrete wall and an extraordinary lime-green lined pool of water. The intense colour of the pool has been changed from a previous, but similarly artificial, pink that was not sufficiently stable in the bright sunlight. It is unexpected and hints at the lecture hall beneath, yet alludes to a series of goldfish ponds that surround the adjacent Polytechnikum by Gottfried Semper (1858–64) as well as the Karl Moser building itself. The north face of Semper's Polytechnikum is dressed with a lyrical blue and white mural, which is similarly unexpected. The mural is clearly expressed as a decorative skin that adds a layer of meaning to the surface, the painted surface being a technique also employed by Gigon/Guyer in a number of apartment projects in the city, albeit as pure applied colour rather than a decorative pattern.

The power of the edge

WHILE the tinted concrete buildings, such as Römerholz and the Stellwerk, use a through colour, several other projects utilize colour that is deliberately expressed as a thin decorative film on the surface. The mass concrete constructions, which are invariably simple rectilinear forms, are then coated with a mineral paint, leaving bare concrete exposed in places to reinforce the idea of the coating. At Broëlberg 1, Kilchberg (1996), the original dark brown (now brown-grey) meets the intense orange at the corner. The conceptual difference between public exterior looking

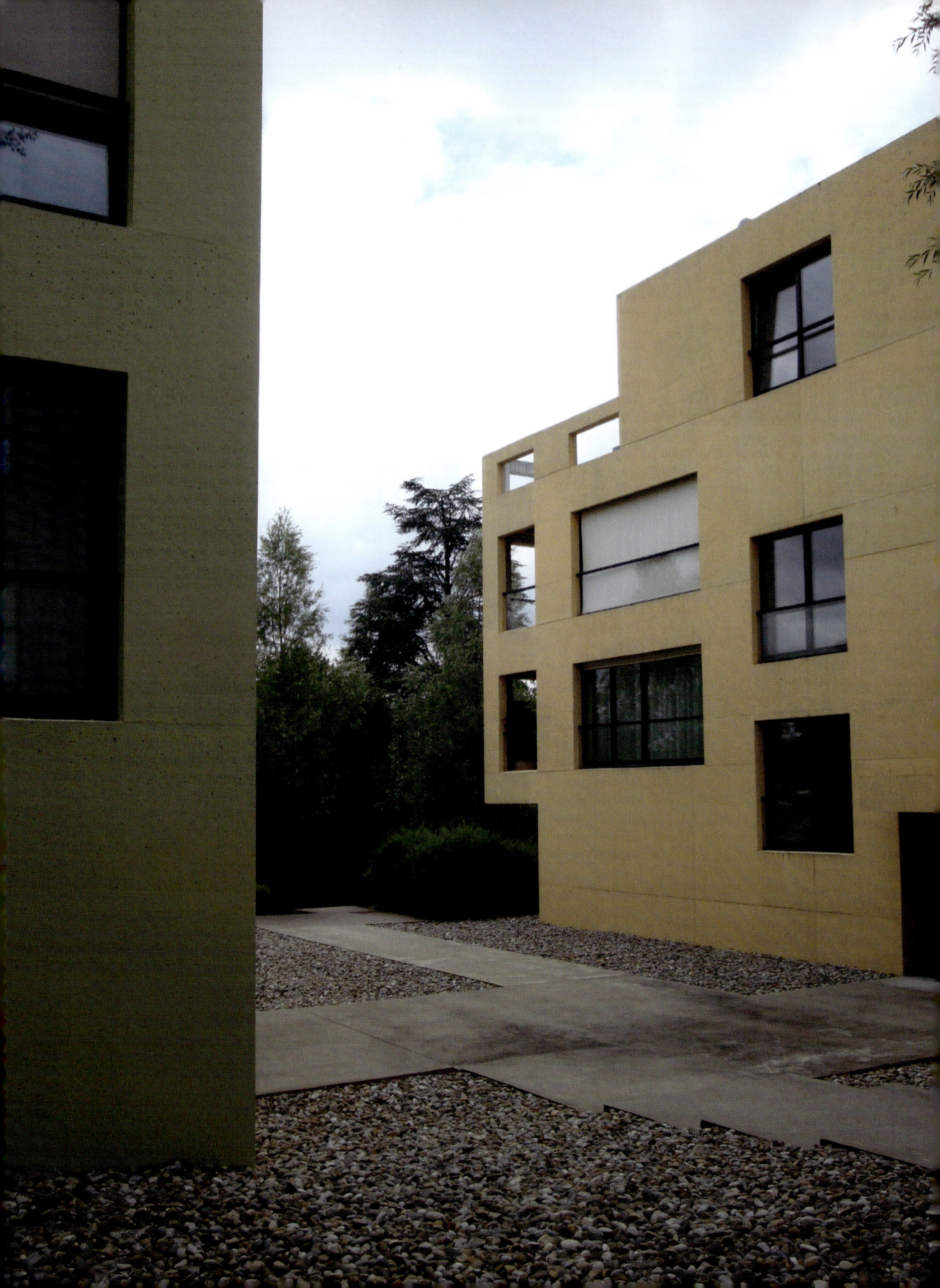

outwards and semi-private, shared courtyard looking inward is explicitly expressed at this point of contact between the planes, where the colours appear to intensify each other.

For a later development of apartment blocks at Susenbergstrasse (2000), high up on the Zurichberg, they worked again with Adrian Schiess. The immediate vicinity is characterized by large single villas along meandering roads that hug the contours. As a response, Gigon/Guyer broke their project into three discrete blocks, using colour to differentiate and to give a sense of depth to the shared public space between. The separation allows each flat to have views in all directions. They considered expressing the difference by using different materials, such as timber, brick, etc., but the cost would have been prohibitive. In the original competition model, the colours are taken from natural features of the site – yellow sunflowers, green and brown, from the woodland and fields behind, with each block adopting a single colour on all faces. The volumes are clear and simple. The architects then approached Adrian Schiess and, working together, the conceptual use of colour shifted to pale pastel shades of warm orange, yellow-green and grey colours applied using a very matt, powdery mineral paint. Curiously, one single facade, facing north towards a group of trees was treated differently, interrupting the otherwise simple expression of volumes. This single facade was proposed to be in pinky red, but changed to blue after the client suggested that, in predominately Protestant Zurich, it was felt to be too close to a 'cardinal red'. The vertical edge at the external corner becomes highly significant with opposing hues – soft orange and blue – meeting at the corner. This facade is the least visible but most surprising and the blue colour appears to push it into the shadows.

At Susenbergstrasse, the colour is taken around window recesses and balconies, reinforcing a sense of depth, volume and mass. In a larger housing project, at Pflegi-Areal in Zurich (2002), colour is used as a purely surface treatment, employed in a manner which could be considered ambiguous. The use of colour brings an additional dimension to the project. The external facades facing the streets, the walls facing inwards to a courtyard, back and fronts are similar in construction and form, but accentuated as being different in meaning through colour. At Pflegi-Areal, depending on the weather conditions, the afternoon sun can reflect a pale peppermint-green light from the inner face of the courtyard onto the surface of the white-painted block to the east. The easternmost facade, facing onto what was an existing mature garden when the site was a hospital, is painted a surprisingly strong blue as a backdrop to the mature trees. It is very common in Switzerland to find highly restrained Modernist architecture immediately adjacent to traditional buildings. The contrast between elaborate and undecorated surfaces tends to heighten the appreciation of each. In the streets surrounding Pflegi-Areal, traditional buildings are rendered and painted in highly colourful, blues, terracotta and greens. The colour, therefore is not

Apartment blocks at Pflegi-Areal, Zurich,
Gigon/Guyer Architects (2002)

facing page
Apartment blocks on Susenbergstrasse, Zurich,
Gigon/Guyer Architects (2000)

Apartment blocks at Pflegi-Areal, detail, Zurich, Gigon/Guyer Architects (2002)

alien in this instance, and the bare unadorned concrete of the west facade and its simplicity could be seen as the most controversial. The way in which the colour is used to define spatial hierarchies within the site is, however, very different from its neighbours. The inner courtyard at Pflegi-Areal is expressed and softened by the green colour, and by a series of extraordinary tree pots dotted across the raised podium. Moving through the spaces, the corners become key, apparently differentiating one space from the next in mid-air. The act of walking through between the buildings connects the spaces. In this project, the colour seems to have a direct effect on the feeling of the spaces, not purely from a visual sense. There are spatial hierarchies and clearly felt thresholds between public and semi-public as one moves through the sequence of spaces. There is a strict sense of control and order to the courtyard that does not invite informal clutter, which has been welcomed at Brunnenhof. By contrast to the raised courtyard, the blue facade adjacent to the shared park accentuates its difference. The colour seems to define an enclosure where none actually exists. The blue wall appears to balance the space, which is lined to one side with tall, mature trees. This space is highly memorable and very different in feel from the courtyard. The materials are the same, but the intensity of this coloured vertical plane together with the landscape seems to define and contain a three-dimensional void, which is effectively without physical boundary. Perhaps this, then, is evidence of a phenomenological use of colour. The body feels boundaries where none exist. The blurring of the volumes into planes by applied colour contributes to the ambiguity in the legibility of these projects. As Max Wechsler notes in his essay 'Beauty is admissible: architecture as visual event', however, the ambiguity in Gigon/Guyer's work is 'quite different from uncertainty or lack of clarity'.[17]

Wechsler argues that the collaboration between Gigon/Guyer and the artists is very different from common 'art and architecture' models where the art is seen as independent of the construction. In these projects art and architecture are, in effect, inseparable. The artists, for their part, are able to work on a large scale and the architects see their work as a spatial framework for art. Neither party can claim artistic authority; their works are true collaborations. Harald F. Müller is quoted as thinking of the colour to be 'like the structural design, part of the building'.[18]

In Gigon/Guyer's case their interest in colour initially also arose from their interest in art. Colour is seen by them as either necessary or not, depending on the specific project. They make that initial decision, but the palette of colours used in their work is then highly dependent on their collaborations and varies with each artist. Max Wechsler invokes a comment attributed to the painter Fernand Léger that architects are simply not competent to decide on the use of colours and from Gigon/Guyer's perspective they admit that they have looked to artists to 'guide them through the colour cosmos … in the spirit of thinking together, searching together, looking together and creating the architecture together'.[19] Clear, sometimes glossy and intensely manufactured colours come from the influence of Schiess and are evident in his own work; earthy, rich and soft with Müller and, more recently, a new collaboration with Pierre-Andre Ferrand has produced very dark shades of 'curry colour', olive green and a strong contrasting yellow at Wädenswil, south of Zurich. Gigon reflects that they do also use colour without the involvement of artists – the gallery at Mouans Sartoux (2004) in France being an example. Although, like most of the architects interviewed for this book, they are not conscious of having a specific colour palette, Gigon notes that a yellow-green does keep recurring and they are somehow drawn to it.

The projects, whether collaborative or not, are confident and assertive, rather than aggressive. In her book *Art and Architecture: A Place Between*, Jane Rendell suggests that the varied output of practices, such as Foreign Office Architects (FOA), Herzog & de Meuron or Gigon/Guyer is evidence of architects seeking to establish a clear identity in the marketplace. If this notion of practice identity is applied to Gigon/Guyer, one can see the use of strong colour as being highly memorable, but it is clearly only a part of their poetic approach to materials, surfaces and an expression of place. Perhaps what sets Gigon/Guyer apart in their use of colour is their openness in relation to ideas and interpretation. The collaborations with artists allow influences from these consultants to infect and direct their work in different directions. The use of colour, although assertive, is often ambiguous in relation to its meaning. It is the fact that surfaces are intensely coloured and, in being so, influence the experience of the spaces and the reading of the volumes that seems to be the point of interest, rather than being dependent on a precise colour for meaning. Their architecture accepts, even embraces, change, both in the naturally ephemeral qualities of colour as light dependent, and change in the materials due to the effects of time. Their work therefore presents an interesting paradox between a tightly defined, rigorous architecture and a loose fit, relaxed manner in terms of its appropriation and interpretation.

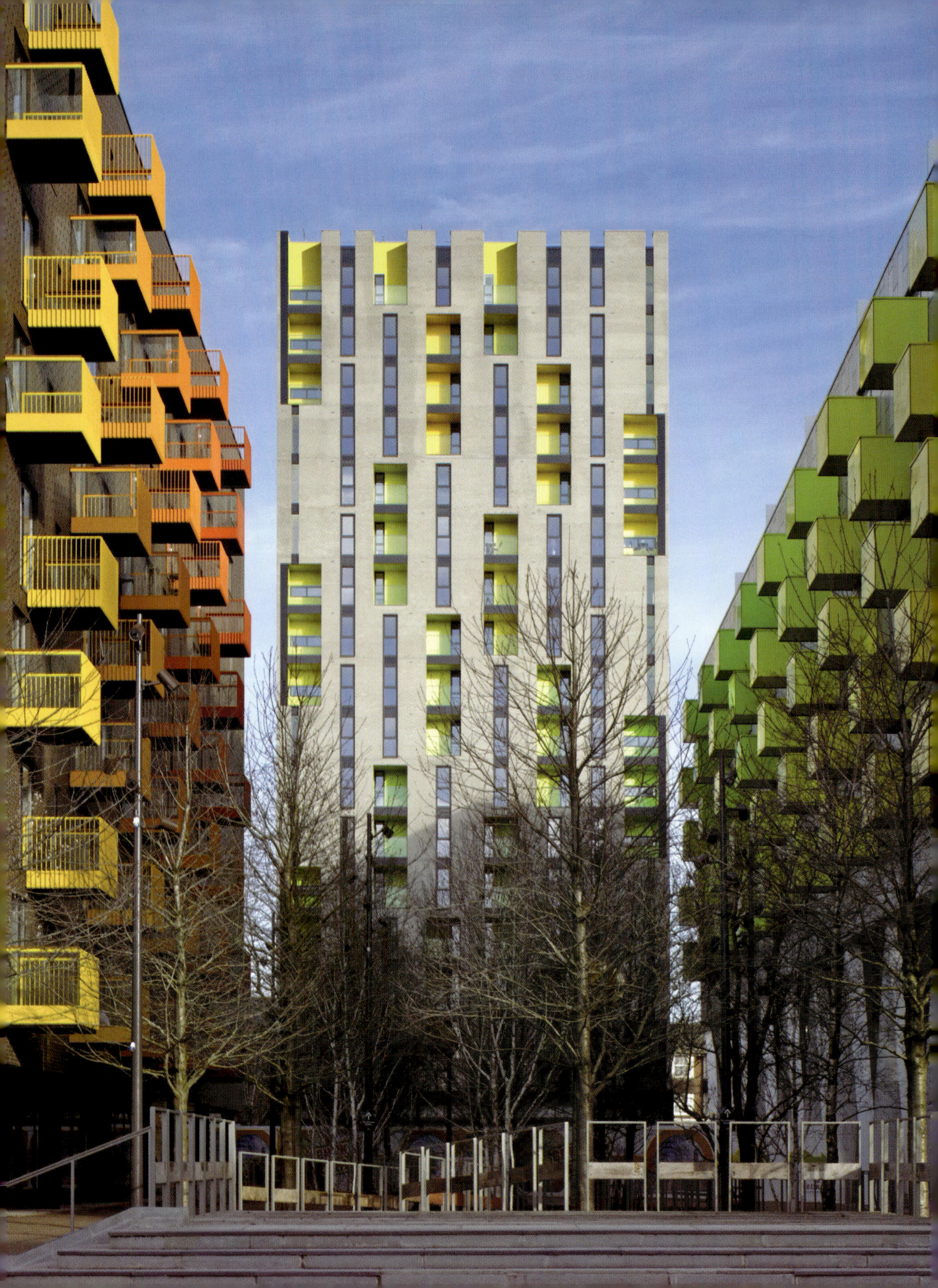

– 8 –

Memories, associations and the brightness of yellow: AHMM

Colour in the urban realm

ALLFORD HALL MONAGHAN MORRIS (AHMM) is an architectural practice whose work is located predominately in London. Their work is rooted in a sense of pragmatic solutions, used not just to solve problems, but to spark further creative ideas wrought out of comparatively low budgets. Building elements are frequently both functional and decorative. Their use of colour evolved in the early days in practice, when it was seen as effectively cost-free and able to define spatial ideas, give identity and reinforce the conceptual understanding of the building.[1] Subsequently, as the practice has become well-established, their use of colour has developed an identifiable set of principles. Their work in the urban realm provides insights into the role of colour in the multi-dimensional environment of the city. The ubiquitous peppering of global marketing, through which very specific hues become firmly associated with specific brands (such as Coca-Cola red) is one significant aspect of the use of colour in contemporary urban environments. Used as part of wayfinding strategies, colour is also fundamental to navigation within complex visual environments. AHMM have used such contrasts to develop a form of branding of their own architecture and on behalf of their clients. In such environments, intense hues serve to contrast with the relatively bland, grey or soft background colours of the contemporary city.

AHMM's buildings are often very simple in form, using uncomplicated, geometrical shapes, such as the primary school at Notley Green School, Essex (1999), which is triangular in plan, and the rectangular Westminster Academy, a high school in West London (2007). This simplicity is explained, in part, by their economically driven principle of trying to minimize external area and maximize useful floor space. They argue that, by doing so, the budget can be stretched to allow for much higher specification of key elements, such as high-quality furniture. The external skin of these elemental shapes is then coloured in a way reminiscent of Wassily Kandinsky's identification of primary colour with primary form. The surface is frequently flat with colour applied as a layer, such as the faience ceramic tiles used

facing page
Mixed-use development at Barking Central, Essex,
AHMM (2009)

in an office development at Crown Street, Leeds (2005), and at Westminster Academy. Occasionally, AHMM will add a layer of complexity within this otherwise reductive approach, by extruding elements of the facade to express depth. Balconies, fins and other coloured elements are introduced to suggest a reading of 'anti-flatness', such as at the Jubilee Primary School, Brixton, London (2002),[2] a large urban mixed-use development at Barking Central, Essex (2007–09), and the housing development Adelaide Wharf, adjacent to the Regent Canal in Hackney, North London (2007). Simon Allford believes that modern methods of construction have a tendency to flatten the skin, whereas traditional architecture celebrates layers, depth and shadows. AHMM therefore use modelling of elements specifically to emphasize this sense of depth. Allford acknowledges a direct artistic influence on the practice's projects from the work of Donald Judd and, in particular, his explorations of surface, reveal, depth, volume and colour. AHMM's aim is to start with something apparently simple and then add layers of complexity and meaning. Colour is used as part of this process, frequently reinforced by applied graphic design.

'Signing' rather than signage

THE simple forms and planar surfaces used as a basis of AHMM's architecture lend themselves to the application of supergraphics, which modify the effect of the volume and contribute directly to the legibility of the spaces and their associated meanings. The concept of signing generally implies a means of communication beyond text, such as that used by the deaf. Signing is a visual language based on physical movements, but can also be a method of tracing meaning through marks, figures and symbols. The use of colour signing in the work of AHMM springs from a desire to explain form and movement and indicate patterns of use. Over time, this sign language has also developed into a very literal use of signage in collaboration with the graphic designer Morag Myerscough of Studio Myerscough. The practice has collaborated with Studio Myerscough for 18 years, starting with a temporary hoarding in the Corn Exchange in London (1995), which was formed in a zig-zag pattern to enable contrasting readings of pictorial images and text from opposite directions

Temporary hoarding at the Corn Exchange, London,
AHMM in collaboration with Morag Myerscough (1995)

Kentish Town Health Centre, AHMM
in collaboration with Morag Myerscough (2009)

in a narrow street. The relationship has become so symbiotic that in some of the buildings Monaghan acknowledges it is 'almost as though our work is like a backdrop for her graphics'.[3]

Studies of how people orientate themselves in space through cognitive mapping, equate a sense of knowing where one is with an ability to conjure up a mental map of spaces. Kevin Lynch's (1960) *Image of the City* and David Canter's (1977) *Psychology of Space* make use of such mental maps. Lynch defines a legible city as one 'whose districts or landmarks or pathways are easily identifiable and are easily grouped into an overall pattern'.[4] The legibility of AHMM's architecture is of paramount importance to the practice, and Myerscough's coloured graphics are a core element in this reading, particularly in public buildings such as Westminster Academy and Kentish Town Health Centre, London (2009). It is significant, however, that Myerscough does not consider the use of applied graphics as a wayfinding strategy, although they do provide implicit identification of routes, corners and key positions within the circulation. She is more concerned with the mental associations invoked by surface

Interior atrium of the Judge Institute, Cambridge University, England, John Outram Architects (1995)

decoration. The intensity of the colours chosen for the supersized images and text is deliberately upbeat and joyful in settings that might be stressful. Kenneth Boulding discusses the role of images, 'built up as a result of all past experience of the possessor of the image', and sees the way in which people store and recall such mental images as a tool in understanding human behaviour.[5] Similarly, Myerscough's graphic images can be seen as tapping into this collective reservoir of memories.

At the scale of the city, this communal social memory is part of a continuum that firmly differentiates the specificity of place. Signage, or surface scripting, although highly intrusive in some situations, is seen as ephemeral, superficial and easily erased. Taken as an integral part of visual culture, Ella Chmielewska argues that it cannot be understood, therefore, as mere surface display:

> Despite its surface position, signage is four-dimensional: it has considerable depth as it is enmeshed in informational infrastructures which are made visible ... and its temporal dimension is embedded in the roles, conventions, and codes governing its use and legibility.[6]

The architect John Outram has written, scathingly, that contemporary architects are embedded 'in an iconically illiterate design culture'. Berating the lack of cultural literacy and the scrupulous avoidance of decoration by the majority of architects, Outram argues that 'the century past was a theatre unfit for human occupation. Its "truths" of naked wood and bare cement, were banal to the point of idiocy.' He therefore interprets the present lack of surface decoration in architecture 'as an imperative to say nothing, to be mute'.[7] He throws down a gauntlet to the architectural profession, challenging it to accept surface decoration as symbolic, as iconographic and essential to architecture, if it is to succeed in engaging with humanity as a multi-dimensional stimulus. His work is polychromatic and each colour is chosen for its role in the narrative. Each is derived to support an architecture, which, to Outram, is simultaneously new and old. Outram's own buildings are a product of this narrative underpinning. Fat, round columns painted earthy brick-red or turquoise, rise to celestial blue ceilings. The most notable example is the Duncan Hall, the Computational Engineering faculty at Rice University, Texas, USA (1996), with its extraordinary pictorial ceiling. The vaulted surface is scripted with iconography to represent a cosmological myth entitled 'The Birth of Consciousness'.[8] The ceiling is itself a marriage of old and new technology, the intricate pattern originated as an A1-sized painting by Outram, was scanned, enlarged on computer, then tessellated, printed using a large-scale printer on to vinyl rolls and then adhered in place in two days. His Judge Institute at Cambridge University, UK (1995), uses a similar palette of terracotta, turquoise and black throughout a soaring atrium rising to a patterned ceiling. The buildings are uniquely Outram in origin, but can also be understood as part of the classical Postmodernism of the 1980s. Outram continues to be astonished that architects choose colour arbitrarily and almost invariably without consideration of potential cultural associations in the mind of the observer.

Outram's work provokes architects and designers to embrace the role that colour plays in triggering more than a simple registration of hue, brightness and saturation. Each tone conjours up associations that may be of cultural significance, have psychological meaning and may be therapeutic or evocative of highly personal memories. The filmmaker Derek Jarman's poetic book *Chroma*, written as illness and loss of sight overwhelmed him, lists many such memories. Each shade of white, brown, green, and so forth, is described not with an image, but purely by words. The text has similarities with Abraham Gottlob Werner and Patrick Syme's (1814) *Nomenclature of Colours: With Additions, Arranged so as to Render it Highly Useful to the Arts and Sciences. Annexed to which are Examples Selected from Well-Known Objects in the Animal, Vegetable, and Mineral Kingdoms*, which was one of the earliest attempts to establish a vocabulary of colour shades by association. Syme's work as a botanical illustrator contributed to the lexicon terms such as apple-green, 'being emerald green mixed with a little greyish white', and pansy purple, 'indigo blue with carmine red, and a slight tinge of raven black', leading to a total of 108 defined shades. Later, Paul Kay and Brent Berlin's (1969) *Basic Color Terms* furthered the search for an appropriate vocabulary of colour, extending the lists to include different terms used internationally to describe colour. Perhaps the coded specification systems that are used by contemporary architects and designers, and which are stripped of such associative meanings, invite a rational precision as opposed to an emotional language of colour. One notes the approach of the paint industry, which employs very different nomenclatures in marketing the same paint to the general public. In literature aimed at a lay audience, for example, ICI/Dulux identify their colour code 30YR 15/550 as Volcanic Splash 4, 43YY 69/543 as Banana Dream 2, and 10GY 41/600, refreshingly, as Kiwi Burst 1. The development of colour nomenclatures is evidence of the importance of recall as a form of mental calibration. Clearly, evocative names evoke connotations and sell products, where codifications do not.

Parrots among the pigeons

In the London Borough of Hackney, the context for AHMM's Adelaide Wharf flats, the local council has used orange-painted hoardings to protect flats that have fallen empty and into disrepair. The colour orange is therefore scattered around the surrounding streets and has come to symbolize public ownership of the empty properties. The orange is a kind of corporate identity, but the choice of colour is disconcertingly cheerful, given what it represents. In some streets the speckling of orange dominates the visual environment. The choice

Orange-painted hoardings in Hackney, London, signifying empty properties (2009)

Housing (top) and letterboxes (bottom) at Adelaide Wharf, London, AHMM (2007)

of a palette of yellows and oranges by AHMM for their Adelaide Wharf project, also in Hackney, was in part a reaction to the client's personal dislike of certain colours, notably green, and the fact that, over time, untreated larch cladding will go grey. The sunny palette of the balconies is therefore intended to counteract the potential banality of the low-cost repetitive facade.

In some of AHMM's work the palette has a more direct meaning. A project for a residential tower in Ghana adopts the colours of a 'washed out national flag'.[9] The project at Barking Central was developed at the same time as Westminster Academy, but was based on yellow because it was on the site of R. White's lemonade factory, which had a well-known corporate identity. The colour was used as a means of retaining a collective memory of the site's previous industrial life. The intensely coloured buildings at Barking, Westminster Academy, Adelaide Wharf and the Yellow Building, the headquarters for clothing company Monsoon in London (2008), have now all become identifiable nodes within their neighbourhoods and contribute to the urban identity. At the same

Westminster Academy,
London, AHMM (2007)

time, AHMM have established a form of branding for the clients, which, particularly because of the intensity of the colour, is visually arresting.

The exterior of Westminster Academy is characterized by horizontal bands of ceramic tiles, gradually lightening from dark green to yellow as they ascend floor by floor. One aim was to avoid the typical bland use of a brick base and occasional coloured panel, which is a common response to the budget constraints on such buildings. The location is in Westbourne Green, in West London, which has a reasonable number of trees and greenery, but the site is adjacent to crumbling 1960s concrete blocks and an austere motorway flyover. Green was considered to work well in this environment, providing a symbolic landscape. The practice considered matching colours with strong associations such as 'Wimbledon green', as the tennis ground is relatively close, or 'British racing green'. The combination of yellow and green had been used in a very early flat design for the journalist Jeremy Melvin, and was seen as offering a wide range of shades and is less harsh than a red/blue palette.[10] The building has been described as follows:

> A long, smooth, sharp block decked in clear glass and shiny bands of green and yellow, with extra stripes of brash colour here and there, this trendy building stands out from its drab west London surroundings like a parrot among pigeons.[11]

The building is unconventionally inside out – with a highly colourful exterior and a subdued, monochrome colour scheme internally, using an exposed concrete frame, charcoal carpets and

Supergraphics in main atrium at Westminster Academy, London, AHMM in collaboration with Morag Myerscough (2007)

natural oak linings. The yellows and greens reappear internally on functional elements – on one side of the solar baffles, for example, which are made very simply from plywood doors suspended in the atrium. Sound-absorbent panels in ceilings are coded, giving each floor a slightly different identity. Occasional accent colours of pink and blue are used in the graphics that support the building with a narrative, devised as a response to the Academy's focal discipline of business skills. It is not, according to Myerscough, a directional wayfinding strategy as such, more a method of recognizing one space over another: 'Once a student gets beyond their first day, they do not require directions everywhere. The main thing they need to know is the door numbers, which were made very prominent and clear.' The project architect adds, 'The colour seems to give an order and clarity to the overall design, giving the spaces a mature yet playful nature'.[12]

The intensely stratified colour of the ceramic-tiled exterior at Westminster Academy establishes a strong identity and a sense of belonging, an extension of the effect engendered by school uniform colours. The external colours are permanent, rooted in the choice of the ceramic material. The branding is part of a deliberate strategy, in tune with the academy's specialism in business skills. The colour of the architecture is now entirely integral to the identity of the school and has become symbolic of its re-invention. While sceptics might say that any new building with such facilities should promote improvement through investment in the lives of its users, AHMM argue that the quality of the architecture of the new school has had a positive effect on the performance and achievements of the students. The combination of architecture and design gives the school a specific character, unlike much of the bland architecture that has been a feature of much of the product of UK school building, as Paul Monaghan has observed: 'Painting a wall orange isn't sufficient'.[13]

Internally, spatial organizations are generally straightforward and comprehensible, and Morag Myerscough's applied graphics strongly support the architects' narrative. The colours were chosen by AHMM in collaboration with her, and tested through a series of large-scale samples in lighting conditions that matched as closely as possible those of the actual building. Digital and physical models were used to experiment with colour and to place the graphics. The designers emphasize the importance of reaching consensus through discussion within a team that included the client. The graphics dominate the interior and add a layer of meaning, picking up on functions such as the dining hall, which is lined with names of the students' favourite foods, to short quotations about learning written along corridors.[14] Romedi Passini has demonstrated, through diagrammatic representation and dissection, the numerous decisions that confront the simplest of navigations undertaken daily.[15] Being able to process visual cues quickly can be seen to relieve stress in building users. This does not directly equate to simplicity, however, as complex pathways can still be legible through the ways in which sensory cues can be extracted. Passini classified signs into three main types: *Directional*, *Identification* and *Reassurance*.[16] The partnership between Studio Myerscough and AHMM have produced notable buildings that are popular with users and can be seen as using graphics and colour as a way of signing to meet all three of these classifications.

Developing the professional palette

AHMM tends to work on a project-by-project basis, deliberately upsetting any sense of consistency in its choice of colour palette. Each sequential project is likely to react against its predecessors, with an explicit desire for difference, rather than similarity. The use of colour is similarly reactive. Having used strong colour on one project, such as their housing at Adelaide Wharf, London (2007), the practice has subsequently experimented with a monochrome, black-based palette for a commercial project at the Angel Islington, London (2010). The black contrasts with a line of mature, protected trees, which are reflected in the matt black glass. The Unity Tower in Liverpool (2008) is also predominately monochrome with subtle shades of green, blue and mauve. At Barking Central, Essex (2007–09), a large multi-use development, a yellow/orange palette is used, varying from the intense yellows of AHMM's hotel building to their final building in the project, which is pure white. Paul Monaghan has commented that they had 'run out of steam on colour' by the final building and, ironically, that it may 'stand the test of time better than its highly coloured neighbours'.[17] This self-critical reflection is clearly part of any robust design process. AHMM proceed by having each project adopt some of their underlying principles of simple form and functional approach, but then searching for its own identity. One senses that, while their use of colour is bold, they are wary of being typecast as architects who always use strong colour and one consistent methodology.

Perhaps this is where AHMM's collaborations with a series of artists and graphic designers serve to stimulate and instill difference. Both the Kentish Town Health Centre and Westminster Academy were shortlisted for the UK's major Stirling Architecture Prize and are examples of collaboration with Studio Myerscough. As with Caruso St John, Steidle Architekten and Gigon/Guyer, their relationship with artists is seen as a productive partnership of equal authority, but there is also a sense that the collaborations are supportive, emboldening the approach to colour by sharing the responsibility for decisions. Allford notes that he feels 'almost guilty' sometimes in trying to justify particular use of colour to himself intellectually, or to a client, through a particular theoretical position. More often the practice sees colour emerge as part of a conversation among a group of designers and would resist the idea that colour choice must be defined by theory, even though they are conscious of the work of colour theorists such as Josef Albers. The involvement of artists adds a level of justification to their conceptual use of colour, to the point where the author is not seen as a single architect, but rather a team of individuals, contributing differing priorities to the evolution of the design. The emphasis on team working is at the heart of the practice, and has been an established principle since the early days of the four founding partners.

The artist Charlotte Ingle worked with AHMM on Templar House – low-cost apartments in Harrow, Middlesex (2005). Here, a very wide palette of colours is used, modifying in sections along the block and across the facade. The colours were derived from photographs that Ingle took of the surrounding context, which she then pixelated in order to abstract a palette. One could see this as a modification of the contextual studies of the French colourists Jean-Phillipe and Dominique Lenclos, who documented a *Geography of Color* from a lengthy analytical process of examining both naturally occurring materials, including soil and vegetation, and building materials.[18] Their factual

Social housing at Templar House, South Harrow, Middlesex, England, AHMM in collaboration with Charlotte Ingle (2005)

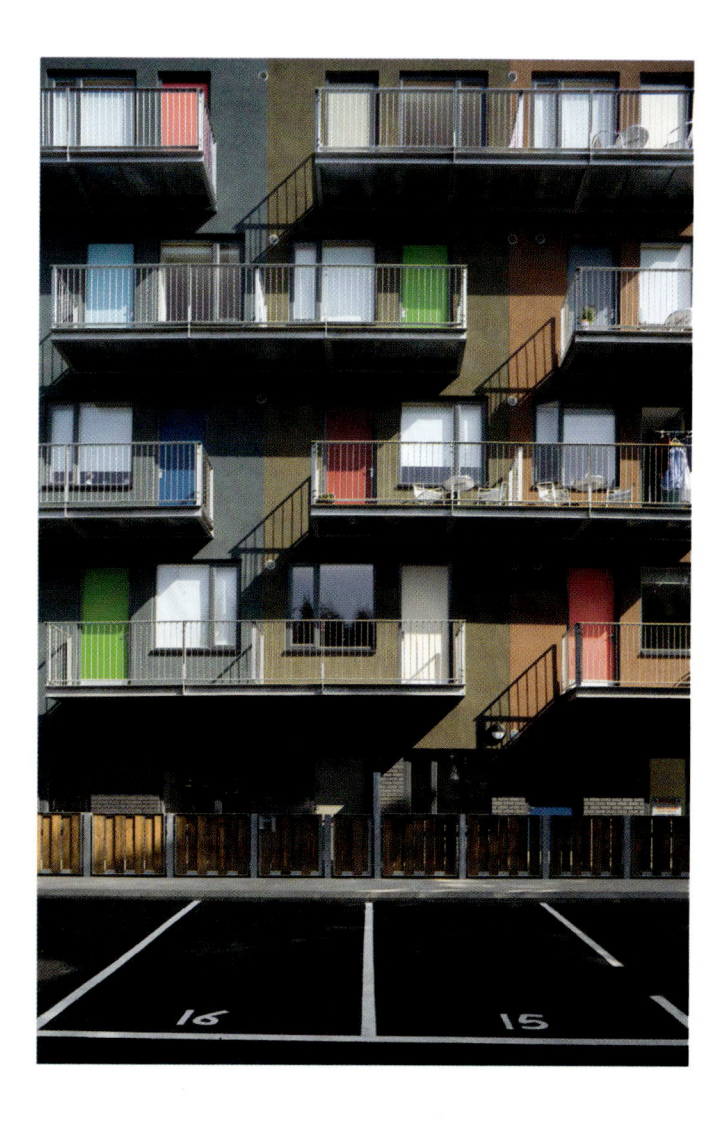

analysis is undertaken using paint scrapes, documentation and observation with the original intention of harmonizing new architectural insertions with the region. Ingle's approach suggests a similar, culturally derived palette, but with a culture rooted in the ad hoc, uncontrolled application of colour in an urban environment. However abstractly derived, the use of such a polychromatic, almost chaotic palette, was part of a deliberate intention to disguise and accept the inevitable individualization that would occur once the flats were occupied.

The mass production process for AHMM's social housing at Raines Court in North London (2003), using prefabricated pods, is subverted by a myriad of soft green/blue/yellow colours. These become more intense as they ascend the building in an attempt to suggest the individual within the regularity of the simple form, and break down the repetition of elements. In this case, the actual colours are less significant to Allford, and the idea of different colours representing different people is the key. If tenants subsequently alter the colours, the architecture is seen as strong enough to absorb this change. By comparison, the use of a single yellow at the Monsoon Headquarters (2008) makes the choice very significant and the colour is protected by being encased by glass, making it integral to the brand image of the building in a far more permanent manner than at Templar House or Raines Court. The conceptual application of colour in the residential projects, therefore, appears to differ from its use in commercial buildings; one accepting change, the other actively preventing it.

The brightness of yellow

MOST architects are not aware of having a specific palette. Despite this flexible approach to the meaning of colour, is there a sense of seriality within the work of AHMM? Six of AHHM's recent projects are documented as a palette arranged chronologically, vertically from top to bottom. Compared with the highly restricted range used by O'Donnell + Tuomey, a wide variety is evident. Yet one can see an underlying consistency in the brightness or freshness of the particular hues and an increasing intensity. AHMM consider that using any colour has an element of risk, taking them out of their 'comfort zone', but it is a risk they appear to relish. AHHM's palette is clearly based on manufactured colours, it is not meant to look natural, and is used to heighten the position of the built object set against either nature, or within blander urban environments. The camouflage blue and grey chequerboard of their flats at Dalston Lane, London (1999), for instance, was highly controversial and caused local outcry when first built, with demands that the facades should be repainted. The most frequent ranges used by the practice are orange to yellow and yellow to green. These are colours that cannot be ignored, they shout loudly, demand attention and become signposts in their urban environment.

Allford has noted that they are interested in the 'brightness of yellow'.[19] Yellow, to Allford, has the ability to lift the spirit, invigorate and engage. He compares unexpected and cheerful colour, in reveals and in the depth of a building, to a vibrant lining of an otherwise safe grey suit. Johann Wolfgang von Goethe and, later, Johannes Itten used the different luminosities associated with elemental colour to demonstrate that a visual balance can be achieved by the asymmetrical combination of colours of differing luminosity. For example yellow, as the brightest colour in our perception, can be balanced by only a quarter of the surface area of its complementary colour – violet. Dynamic equilibrium, as used in elevations by Adolf Loos and Charles Rennie Mackintosh in relation to the placing of irregularly spaced windows, can equally be applied to the distribution of coloured planes relative to their respective areas.

facing page
Colour palette of architects AHMM as compiled by author.
Six projects are shown chronologically down the page from the soft hues of the social housing at Raines Court (1999) and Templar House (2005), to the clear, strong yellows, oranges, greens and blues of Crown Street (2005), Westminster Academy (2007) and Adelaide Wharf (2007), and a return to milky shades at The Unity (2008) in the lowest line

Johannes Itten provoked a leap in our understanding of colour balance in relation to the quantity of particular colours and their relative intensity.[20] His simple rectangular diagrams serve to illustrate mathematical relationships that had been documented by Goethe. Itten enforces the mathematical analogy in defining this 'contrast of extension' in the form of an equation:

force equals brilliance plus extent.

On Goethe's simple scale of intensity, yellow is 9, orange 8, red 6, violet 3, blue 4, green is 6. Itten gives fractional relationships as proportions, for example, yellow:violet 9:3 = ¾ to ¼, to calibrate the effect of the relative intensities.

Yellow appears, therefore, three times as intense as violet, even though they are seen as opposing. As colour is not intrinsic to the object, but is a product of the way in which light is either reflected or absorbed by the surface of the object. We see yellow as brighter because the cones in our eyes are more sensitive to yellow and less so to violet, and so we register it as being more intense.[21]

Johannes Itten's (1963) diagram of the quantitative mathematical relationships of contrasting colours in order to achieve a perceived balance in the eye. Only 1/4 of the area of yellow will balance 3/4 violet; red and green are equally bright, whereas orange balances blue at 1/3 to 2/3

The 'Rational Color Circle' designed by Faber Birren (1934). Birren argues that the eye can see more warm colours than cool so the position of the balancing grey centre point is shown offset to take account of the unequal brilliance of specific opposite hues.
L = leafgreen and T = turquoise

Both Itten and, later, the colour theorist Faber Birren suggested a distorted irregular model for representing colour, unlike the simple colour 'solids' of Philipp Otto Runge and Isaac Newton. Faber Birren notes 'all color circles are for the most part satisfactory for color harmony purposes,'[22] but his own 'Rational Color Circle' from 1934 is eccentric and unbalanced. His circle has 13 colours grouped with more space given to 'warm' colours – red to yellow – and less to 'cool' – green to violet – to take account of the relative brightness and the ability of the eye to distinguish a greater range of warm colours. Thickly drawn lines across the circle connect four primary opposing colours, yellow to blue, green to red, and a grey node is then placed off-centre inside the circle to denote the focal point of this imbalance. This is a very subtle diagram, which aims to represent the stronger intensity of warm colors. Similarly, the distortion evident in Itten's colour diagram is reminiscent of the imploded map of Europe drawn by OMA[23] to represent the compression in time versus distance

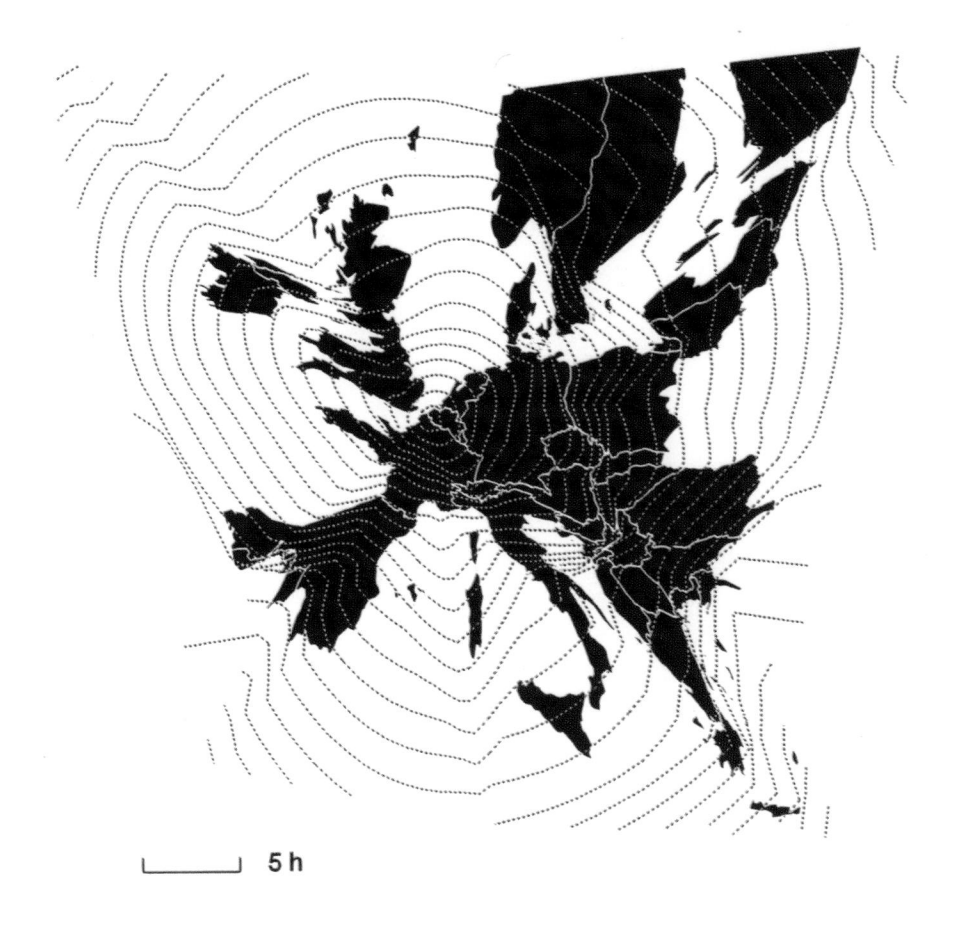

Imploded, distorted diagram of Europe by OMA Architects, Netherlands, to show the reduction in journey times/distances following the introduction of the TGV and the Channel Tunnel (1995)

after the introduction of the Channel Tunnel and high-speed Train à Grande Vitesse (TGV) connections, which suddenly propelled the backwater town of Lille to a new prominence as a central hub of the continent. Irregularity can also be seen in the contemporary representational diagrams, most clearly in the case of the RAL system.[24] The simple colour sphere, as drawn by Runge, has morphed into a distorted form with some portions extruded to articulate the availability of colours with different chromatic levels.

This distorted equilibrium of colour intensity may be instinctive to many of us. We are aware of the brightness of yellow, the vibrancy of orange and the relative calmness of violet. Developing an understanding of dynamic asymmetry in colour is useful in any composition, either in a facade or spatially, to suggest instability in the haptic experience of a space. The readily available tools for colour selection – systems, atlases, fans and computer software – provide us with representational methods in which colours are arranged in regular, ordered sequences.[25] In relation to harmony and dissonance, the order underlying this systematization or indexation of colour is intended to aid the navigation of colours that have a pleasing visual relationship, or that deliberately invoke a visual tension.

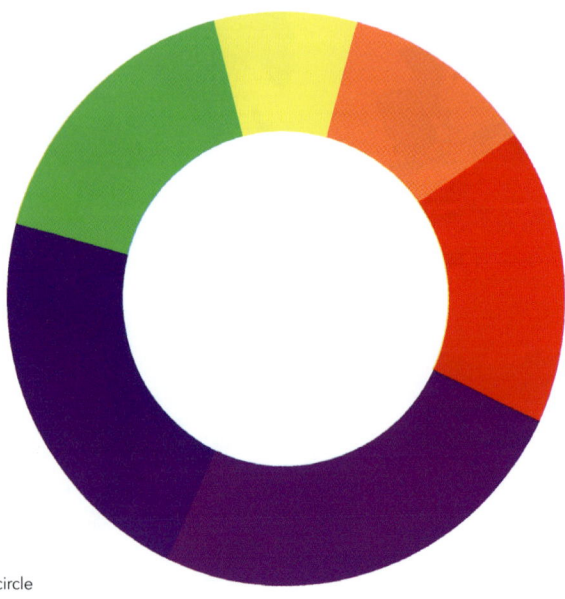

Johannes Itten's colour circle showing the effect of the unequal brilliance of hues – the segments vary to reflect the different area of colour required to give an equal visual effect (1963)

AHMM's approach to colour accords with their underlying architectural doctrine. Their work has been described as being informed by 'principled pragmatism' as opposed to pragmatic principles.[26] Colour is not used as a prosthetic for poor design, it is inherent to an overall philosophy of clarity and unpretentiousness. Particularly where used externally, colours in architecture, in contrast to art, are invariably restricted by the technical specification, such as the need to minimize maintenance. The durable paint specified at Adelaide Wharf is similar to that used in extreme offshore locations and the choice of shades, while still extensive, is not as wide as in other systems. Allford is quick to note that:

> this is not a painting, these are buildings, with maintenance strategies and warranties … so in a sense the practical had a huge impact on the theory … Theory can be a tyranny, it gets bastardized and taken in another direction by production – it has to be a flexible model.[27]

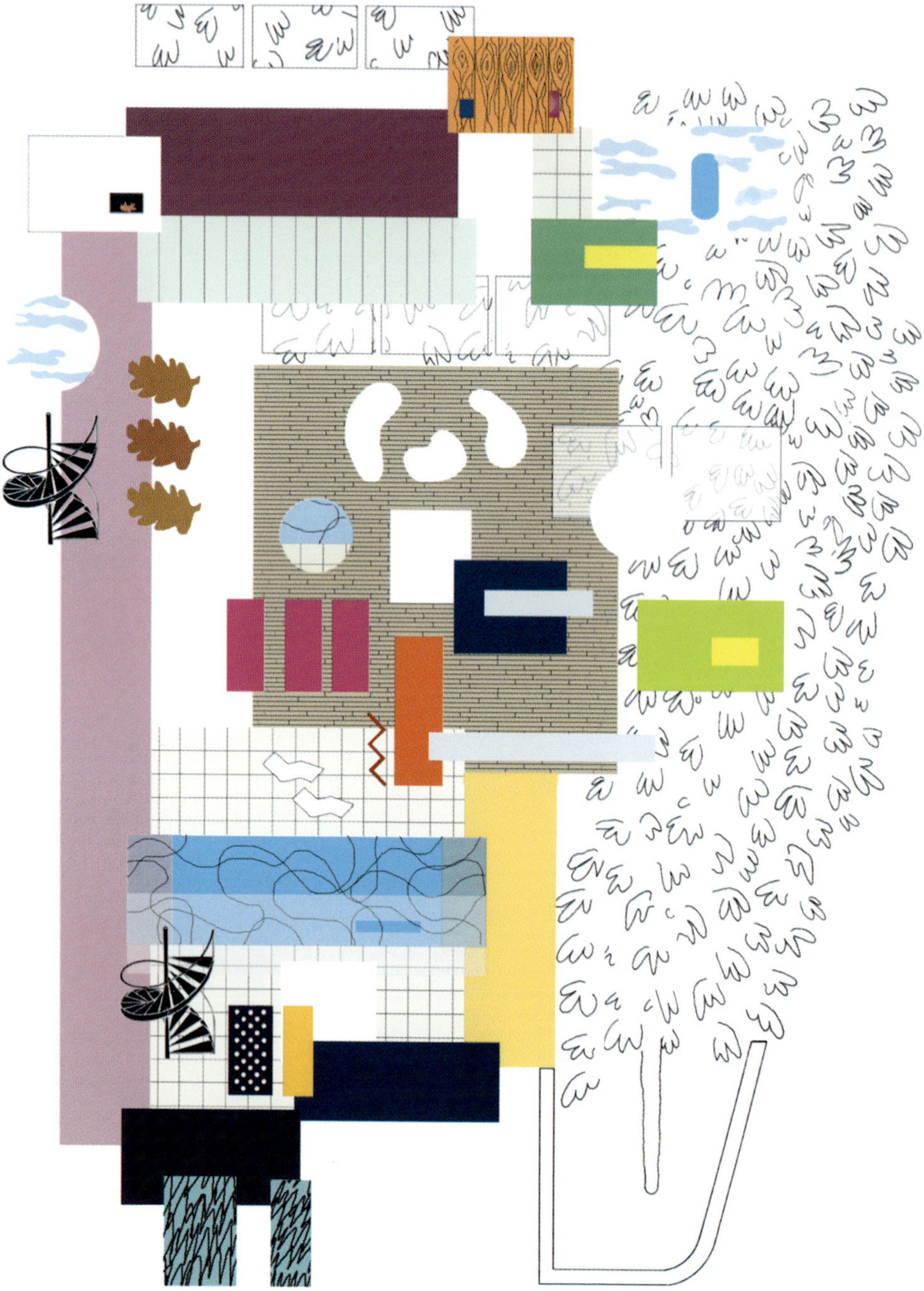

– 9 –
Synergies and discords: Sauerbruch Hutton

Sauerbruch Hutton are perhaps the best known of contemporary architects whose work has become synonymous with their use of colour. They are unusual in that they have a clear understanding of colour theory, which they apply with rigour and through painstaking research and development. Their buildings, such as the Fire and Police Station in Berlin, deliberately use dissonance and simultaneous contrast to heighten the effect of one colour on another. Rarely using single colours, but a combination of a wide range within the same project, recent projects such as the Brandhorst Museum in Munich can be seen as adopting pointillist techniques. The colours become mixed in the eye of the observer, rather than on the surface, creating ambiguity and uncertainty. Sauerbruch Hutton use colour to evoke a sense of well-being. Through their work, this chapter introduces the broad concepts of colour harmony, equilibrium and dissonance drawn from colour theorists such as Goethe, Runge, Itten and Albers.

Between the physical and the visual

'Colour is for us like brick. You would not raise an eyebrow if someone like Louis Kahn designed another brick building. So it is just because colour is being so underused at the moment that colour architecture stands out.' [1]

If there is a single contemporary architectural practice that is firmly associated with the use of colour in the minds of the present generation of architects, it is Sauerbruch Hutton. Their output of polychromatic buildings has become emblematic of their practice, to the extent that one could consider colour *as* the ethos of the office. This would be a naively shallow reading of their work, but such is the strength of the visual imagery that it is certainly the dominant characteristic of their architecture.

facing page
Abstract concept painting by Sauerbruch Hutton
for the H House, London (1995)

Matthias Sauerbruch and Louisa Hutton met while studying at the Architectural Association in London. Matthias is German and worked with Elia Zenghelis and Rem Koolhaas at the Office for Metropolitan Architecture (OMA) in Rotterdam after completing his studies. Hutton also worked for a seminal office in the UK, that of Alison and Peter Smithson. Sauerbruch Hutton have always balanced teaching, practising and writing. These formative years spent working with such influential architects, thinkers and writers clearly set them into a way of working which is highly productive in its output in both buildings and creative thinking. It is unclear whether the clarity and relative simplicity of their architectural principles has been necessitated by the rapid growth of the office, or has been the catalyst for their success. Few offices have such consistency of output and such a recognizable product. Colour, and polychromy in particular, is so closely associated with their work that they may find it difficult not to be typecast. Indeed, some critics simply cannot see beyond the colour being mere surface decoration.[2]

Unlike most of the other architects interviewed for this book, there are a substantial number of texts, interviews and lectures by Sauerbruch Hutton through which their use of colour, specifically, is scrutinized by critics, or where they reflect on their own practice in relation to colour. Most of the other architects give the impression of being in a happy state of uncertainty; to some extent being unsure about exactly why certain colours are used or, in some cases, without dwelling too long on why colour is present. A numbers of others defer to colour consultants, pleased to share the authorship. Being set firmly in the limelight has its disadvantages. Sauerbruch Hutton do not use consultants, and the partners are involved in every project.[3] Being constantly asked to defend their use of colour has meant that they have read, researched and reflected more than most, and argue that the use of colour in Western culture has suffered a long history of prejudice. It is often associated with naivety, gaucheness or a lack of sophistication, whereas white, grey, monochrome or naturally coloured materials are more often considered to be the product of educated minds.

Immediately, then, we have a paradox. Sauerbruch Hutton's work is inseparable from colour, yet if one were to discuss colour alone, one might fall into the trap of crediting colour with an independent existence.[4]

So where does the colour come from, what does it do for their architecture and why do Sauerbruch Hutton keep coming back to it?

In common with a number of contemporary international architectural figures, Sauerbruch Hutton spent their early days post-education, teaching, drawing and painting. Architectural work was scarce in the late 1980s and early 1990s, and paper architecture became the vehicle through which architectural ideas could be developed. It seems apparent that the time spent in two-dimensional studies, paintings and drawings has been seminal in the development of a generation of architects, many of whom trained at the Architectural Association in London during Alvin Boyarski's period as Chairman.[5] When one considers the work of Peter Salter, Peter Wilson and Julia Bolles, Nigel Coates, Rem Koolhaas' founding partners in OMA (Madelon Vreisendorp and Elia Zenghelis), Zaha Hadid and Sauerbruch Hutton, there is patently a link between the investigative drawings and the resultant architecture.

Initially, unable to design buildings, their drawings and paintings explored imagined spaces. The Russian artist El Lissitzky is cited by a number of these architects as being influential, particularly his *Proun* series, which envisages three-dimensional space within the two-dimensional realm. The paintings are often monochrome in structure, but with colour as a dynamic element. El Lissitzky, who never properly explained the title of these paintings, defined them ambiguously as 'the interchange station between painting and architecture'.[6] Zaha Hadid's stunningly beautiful paintings from the early Hong Kong Peak competition in *Planetary Architecture II* folio (1983) draw on the work of El Lissitzky and Kasimir Malevich.

The vital role of the architectural drawing in the production of space has been well-documented by a number of writers, notably by Alberto Gomez-Perez, and for all of these architects (and others such as Steven Holl, Aldo Rossi, Daniel Libeskind, Bernard Tschumi and 'Grazer Schüle' architects Günther Domenig and Szyszkowitz + Kowalski), colour is an integral part of the drawings.[7] The relatively safe space of the drawing and painting allows the mind to develop, clarify and test the project, as well as becoming a vehicle through which to communicate to different audiences. Peter Wilson's *Bridgebuildings + the Shipshape* (1984) explores narratives as generative architectural proposals and, having started to realize the work, the subsequent book with Julia Bolles, *Western Objects Eastern Fields* (1989), provides an intellectual output in writing and drawing through which the conceptual ideas, not always divulged to a client, can be explored.

What is the role of colour within such images?

Mark Wigley, in dissecting Le Corbusier's oscillating relationship with colour, notes that colour was seen as a mode of communication because 'the eye is guided by color signals to certain points of interest' and that, by 1938, Le Corbusier addressed students by recommending that they use colour in their design drawings to 'clarify and disentangle' their ideas.[8]

Louisa Hutton and Matthias Sauerbruch developed a very particular thick line drawing style in which colour is applied in disembodied blocks, slightly offset from the objects. These abstract drawings, which are overlays of accurately drawn perspectives, are seen as a 'condensed outline of space' and are used 'in order to avoid the trap of naturalistic representation'.[9] The deliberate disconnection of the colour blocks suggests a hierarchy within the composition of the drawing as well as in the architectural proposition. The amorphous colour is also loosely related to the zones of movement, landscape and focal points. Their interest in spatial counterpoints was partially derived from English landscape gardens, where form, elements and colour are placed in relationships of foreground and background, all of which contribute to the sense of place.[10]

2D to 3D

The first real testing ground came in a series of residential projects in London. Abstracted drawings, such as those for the H House (1994–95) flatten the perspective completely, reducing the composition back to a two-dimensional representation of coloured fields floating in vague proximity to their three-dimensional hierarchy. The placing of coloured planes aims to create a spatial dialogue between inside and outside space, particularly in the case of a long 'Mexican pink' wall which bounds the property and acts as connector and ordering device, both in the drawing and in the house itself. The client was a film-maker and clearly understood and enjoyed the spatial and visual intent of the short and long view, and that the colour is an integral part of this.[11] The project echoes the work of Luis Barragán (1902–88), particularly in the use of strong pinks, rust and planar elements surrounding and claiming space, inside and outside. Barragán had the luxury of time in being able to specify a grey render and then, only after observing large colour swatches on site, in all weather conditions and light, finally deciding on the specific hue. Sauerbruch Hutton describe the house as a having the spatial logic of a large, inhabited painting. Its architecture creates ambiguity between the physical reality and the visual perception of the space. In addition to these planar fields, colour is also used to define large pieces of built-in furniture that serve to store the apparatus of everyday life. As in Le Corbusier's work, the intense colour defines the element and separates it from its background, but the actual colours were inspired by a trip to India, where they enjoyed the vivid clashes of pinks and reds.

In their earlier L House (1990–92), a conversion of a London terraced property, the colour is given a presence, which to them is no less defining than the physical dimensions of the space. Colour is used to transcend the narrow spaces, using visual effects similar to those employed by Josef Albers, whose paper studies with flat colour suggest an apparent optical depth.[12] By removing all the partitions and replacing the roof on the top floor with glass, the space is flooded with light and feels almost external, with a direct relationship to the sky reminiscent of a James Turrell installation. In this case, Sauerbruch Hutton argue that they used colour as an actual building material, suggesting that the space appears as if constructed out of vivid hues. Kurt Forster interprets this use of colour as 'removed from ornamental intent – anti-decorative'.[13] This approach echoes that of the Bauhaus, which placed colour alongside clay, stone and glass as a material that should be mastered as part of the standard curriculum. As noted in Chapter 2, the de Stijl architects considered colour as an element of construction in a similar interpretation. Colour is, of course, both material and immaterial – physically present and yet entirely a phenomenon of light reflecting from a surface, and hence its complexity.

In common with other architects interviewed for this book, Sauerbruch Hutton consider colour simply as one of the tools available to an architect, most significantly perhaps in relation to how people feel emotionally in and around buildings.

Well-being

THE building for the German Federal Environmental Agency in Dessau (2005) is a good example of Sauerbruch Hutton's use of colour to 'nourish the sensual perception of its users'.[14] Situated on a brownfield site near railways sidings, this very large building was intended as an exemplar of sustainable design. It benefits from features one would expect in an ecologically ambitious programme, such as ground source heat pumps, high levels of insulation and a large atrium. For Sauerbruch Hutton, sustainability is not only a scientific term, it has meaning across three differing scales of application: city, building and human. The use of colour is almost entirely related

German Federal Environmental Agency
in Dessau, Sauerbruch Hutton, (2005)

to emotional well-being, which is difficult to quantify, but which can be derived from a sense of place, and from the aesthetic qualities of the building. Hutton frequently uses the word 'delicious', to describe colour, implying that one can almost taste something which is actually purely visual. A similar connection between visual stimulation and taste is experienced in food presentation. The perception and phenomenology of colour is further discussed in Chapter 6, but it is clear that Sauerbruch Hutton use colour in a haptic sense, to shift the mood of both users and observers. The constantly changing perspective of their amorphous built forms heightens this experiential quality. Internally, the atrium of the Dessau building is on such a scale that it has become a new urban space. Again, there is reference to their interest in the picturesque, in the manipulation of form and colour to enliven the space. But, beyond the visual, they are directly exploring the ability of colour to engender a sense of well-being, wholly appropriate to the environmental programme of this building. This is repeatedly used in their architecture to tune into a human need for colour as a sensual and deeply motivating psycho-physiological mechanism, 'working in this way, a co-existence of the visual, the haptic and the corporeal can be achieved'. Their 'projects insist on a bodily engagement, and the roaming eye seems to be able to touch the surfaces, the spaces'.[15]

The form of the Dessau building, a distorted oval with a tail, snakes across the site, fragmenting the form in order to reduce the apparent mass of such a large, potentially monotonous ribbon of office accommodation. The colour is used on the external skin of the building not only to further embed the building in its context, but also as a joyful, stimulating surface which acknowledges the existing colours of the urban fabric in red-purples of the former industrial brickworks and two different 'families' of greens alongside a park. Each of seven different chromatic themes gives character to the external spaces adjacent to the building, defined by the sinuous form. Colour, form and surface are mutually supportive. The pinks and ochres used in a separate, rendered building pick up on characteristic colours from the fabric of Dessau.

Part of a sustainable architectural policy is the reduction of transportation through a return to locally sourced materials but, in common with most materials and building components, architects have become less likely to source colour locally. Globlization and mass production methods have tended towards pigments and pigmented materials being sourced from only a few internationally operating sources. As a result, architects either choose colours directly from manufacturers' swatches or from more abstracted antecedents in paintings that are then matched to swatches. Few will make a selection based on a close study of a local palette, as advocated by Jean-Phillipe Lenclos.[16] Neither do Sauerbruch Hutton's buildings replicate soft, naturally occurring colours. Their buildings, such as in Dessau and at Sheffield University, England, may take note of observed contextual colour, but then the application remains entirely alien in form, material and in colour. In Sheffield, the brick reds of the surface panels echo a large Victorian maternity hospital that has great significance to the people of Sheffield. The turquoise-greens are used to form a visual gateway in tandem with an adjacent pre-patinated copper roof. In both projects, the textures, reflectances, degree of transparency and translucency of the various cladding elements are woven to give a further layer of variety to the surface when viewed anamorphically.

Subverting the form

An earlier project, the Photonics Institute (1995–98) in the old Soviet Army quarter at Berlin-Adlershof, close to Schönefeld airport, uses colour to subvert the form of the building and structure. The building houses scientists who research the properties of light and, paradoxically, the brief requires the interior to be essentially dark. Their solution was a deep-plan design with light admitted selectively through atria and controlled by perimeter blinds.

The contextual analysis of the site suggested a permeable form that would allow the public to meander from a large park through the site, opening up what was previously a highly secure area of the city as part of an urban strategy to relocate sections of the Humbolt University to the site. Sauerbruch

Photonics Institute, Berlin-Adlershof,
Sauerbruch Hutton (1998);
(below left) atrium columns at the
Photonics Institute

Hutton subdivided the form into two amoebic figures. The resultant space between the objects suggests movement as well as giving a sensuous shape to the profile of the contained sky. The various coloured blinds set inside the facade are intended to reinforce a feeling of instability, of oscillation, of the constantly changing qualities of light. Inside the atria at either end of the building, tall concrete columns are painted with abstract rectangles of colour. Rather than expressing the solidity of these elements, Sauerbruch Hutton drew on the use of colour by Owen Jones on the structure at the Crystal Palace (1851) to express a lightness by subverting the form. Mies van der Rohe used mirrors on cross-shaped columns at his Barcelona Pavilion (1928–29) in a similar optical illusion.

One might not expect scientific buildings to be as colourful and playful, but their pharmacological research laboratory at Biberach (2000–02) uses an abstracted pattern derived from a

sedus

microscopic image to similarly break up the surface of what might otherwise be an anonymous grey box and suggest the optimism of research. The concept is taken further in a huge warehouse for Sedus at Dogern in the Black Forest (2001–03) that uses 20 different coloured ceramic tiles, irregularly distributed across the facade, producing a pixelated surface, which also subverts and dissolves the very large form against its context. Far from becoming mute, however, the surface is so striking that it has become symbolic for the company and is used on their corporate identity.

Irregularity in the facade

IRREGULARITY in the composition of the facade was used by Arts and Crafts architects such as Philip Webb and was a deliberate gesture intended to imply that the artist was not entirely in control of the composition, that the picturesque response to the situation and to internal functions would dictate, for example, the positions of window openings. The architect was therefore to be seen in the more deferential role of interpreter, rather than dictator of form and composition.

In Lucien Kroll's work from the 1970s, such as his buildings for Université Catholique de Louvain in Brussels, he emphasized irregularity, used as an expression of humanity. The haphazard and anarchic composition is designed as a reaction to the mass-produced, soulless buildings of the surrounding campus. The numerous different window types used are a deliberate counter-expression of the regularity of the student rooms behind, used to convey individuality within the constraints of repetition. It is common to see colour used in a similar manner. Many recent architectural facades do not just play with irregularity in relation to composition, but also in the selection and juxtaposition of differently coloured panels, seats, doors, etc. So how do architects make decisions on which colours to select, and how to arrange them?

Tight urban sites, which may restrict an irregular plan form, tend to invite a more playful experimentation with the design of the facades. Hendrik Petrus Berlage's[17] Holland House in London sets up regulating lines ordering the facade and colour gradation vertically over the surface. The facade is highly undulating to create shadow and depth, and the vertical lines taper to the top of the building to give an apparent reduction in the weight. The green-glazed faience also varies to reinforce this effect. David Leatherbarrow and Mohsen Mostafavi suggest that the architect is employing an effect similar to that used by the Impressionists to create shadow and reflectance, which makes the facade appear 'uniform but also unstable'.[18]

In commercial projects, which are a substantial part of Sauerbruch Hutton's work, the facade can be the only element open to the architect to design. Speculative offices, based on a structural grid and fitted out by a specialist contractor to a corporate identity, divorce inside from outside. In these cases, all that remains to be designed may be the wrapping of the box, although a skilled

facing page
Sedus Warehouse, Dogern, Black Forest, Germany,
Sauerbruch Hutton (2003)

Synergies and discords: Sauerbruch Hutton **146**

Excerpt from *4900 Colours*,
Gerard Richter (2011)

architect may still produce a highly sophisticated commentary on the box and its skin.[19] In this context, it is understandable that the designers may introduce compositional rules to give some sense of control and authority.

Whether symbolic of democracy or complexity, the use of irregularity in form, or facade, or both, has become prevalent. As the use of pure chance as a design method for generating architecture is ultimately likely to give an unsatisfactory, or unsettling result, some form of control, either by interpretation or by regulation, is required.

If architects introduce self-regulation in order to control their decisions in relation to the rhythm and composition of facades, is the same true in their determination of colour?

In his designs for the stained glass windows at Cologne cathedral, the artist Gerhard Richter used a computer to generate the random sequences of colours. While the original palette of colours was derived from the medieval colours of the existing stained glass, the pattern is random on one half of the window, and then reflected symmetrically to form the whole design. The windows were designed at the same time as his *4900 Colours* series, which was shown at the Serpentine Gallery in London. A palette of 25 colours, chosen in regular steps of hue, and from light to dark, are numbered and then put in order by a computer. He then made 49 paintings of 100 squares as a sequence. The work is unsettling, partly because it implies an inexhaustible number of configurations, all of them different, but equal, and with no hierarchy of meaning. For the church, the celebration of chance now embedded in the windows, and the apparent lack of obvious meaning, has been controversial.

Richter has used swatches of colour in a series of iterative paintings since 1966. His first paintings made reference to the colour charts of paint companies, evoking them by replicating the white lines between colours. Richter has continued the series, intermittently returning to abstract colour investigations. He has noted 'We can paint anything. But the hardest part is judging whether what we have done is good or bad. Seeing is effectively the determining act which puts the maker and the viewer on the same level.'[20] We have seen a growth in gridded architectural facades that use irregularity both in composition and in the distribution of coloured panels. By what criteria do we judge whether the result is good or bad? Is there a method between rigid prescription and random, subjective and spontaneous selection that is practically useful to contemporary architects?

Many architects, when confronted with choosing colour, admit rather sheepishly to the fact that their choice is arbitrary. They have little means of judging what works and what does not. While one can point to the lack of training and discussion of colour in architectural education as being the source of this problem, the discomfort felt in admitting that they may not be in complete control also has its root in education. As design professionals, we are trained to seek meaning, reason and be able to defend decisions. Richter positively invites chance, and confronts us with uncertainty and ambiguity. For architects, conceding that colour choice, or indeed any design decision, can be arbitrary makes us uncomfortable and embarrassed. We find ourselves in a confused state; we would prefer some rules, some justification for the decision-making process, yet taking the view that colour choice should be prescribed and limited, that there are restrictions on colour use, is equally unpalatable to many architects.

With Gerard Richter, the element of chance is highly controlled within a grid. The grid is the container and regulates chance by giving a physical structure. As with Le Corbusier's use of 'regulating lines' to constrain proportions, or as is evident from investigations into the composition of contemporary irregular facades, where there are no other generating factors, we tend to invent rules or guiding conceptual principles.[21] Rules, similar to those by which we play games, provide order and a sense of the designer's authority over the truly random. Benjamin Buchloh has noted of *4900 Colours*: 'Simultaneously the painter insists on a total confinement (of the grid) and a totally aleatory structure of random chromatic distribution.'

Sauerbruch Hutton's GSW Headquarters building (1991–99) has become similarly symbolic of the regeneration of Berlin. It is a highly significant building and can be considered to have spawned numerous polychromatic derivatives. Many of the recent buildings, which employ random coloured panels across their facades, nonetheless use colour in a fairly one-dimensional and naïve manner. GSW is the most significant building for these architects, taking them from relative obscurity to international prominence. While not attempting to give a full account of the architectural design, which is very well-documented, we will focus on the use of colour in the building, which is integral to its architecture on a number of levels. In urban terms, the architects used both colour and form as a critique of what they saw as a bland urban policy, while at the same time embedding the form

GSW Headquarters, Berlin, Sauerbruch Hutton (1999)

in the grid of the city. The project cleverly incorporates an existing 1950s office tower in a conglomerate design of five symbiotic objects, each with its own identity and colour theme.[22] The strongest body of colour is on the huge expanse of the west facade.

The tones were chosen to complement the typical grey skies of Berlin and also symbolically in relation to the environmental design of the building. The west facade is the receiver of the solar energy and the colour expresses that warmth, supporting the narrative. As with their other work, the GSW Headquarters is designed to incorporate low-energy, sustainable principles. The highly sophisticated facade allows individuals to choose when to rotate the shutters and, in doing so, spreads random patterns across the facade. In choosing the actual tones to be used, some deep colours had to be rejected as they absorbed too much heat. 'The colour and depth of the facade and the movement of the shutters give the building legibility, idiosyncrasy and a certain sensuality.'[23] Hutton compares it to a large-scale dynamic painting. One of the key differences that sets Sauerburch Hutton's work apart from their imitators is the diligence involved in the actual selection of the colours and the composition of the facade. Choosing the colours is an iterative process, made much more complex as the greenish self-colour of the glass exterior skin distorts the colour of the panels. Making the composition, setting the rhythm, proportion, proximities of colour and scale of the panels, as with all of their buildings, requires an extraordinary investment of time by architects, manufacturers and clients. A further skill lies in making such a dynamic facade of irregularly placed coloured panels appear visually balanced.

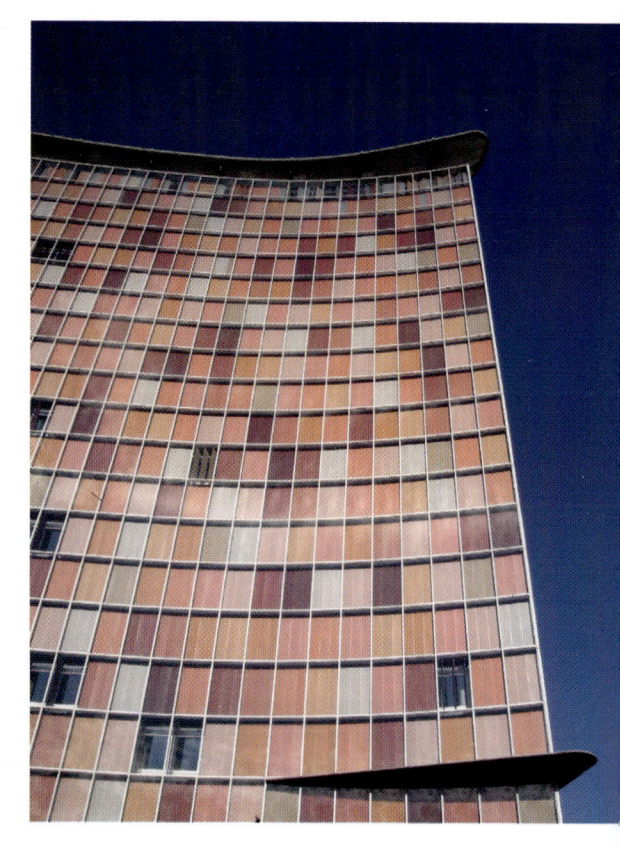

GSW Headquarters, Berlin, detail, Sauerbruch Hutton (1999)

Harmony and dissonance

SINCE the Renaissance, theories of harmony in relation to colour, according to Westland et al.[24] have included common themes: harmony between colours of the same hue which vary in chroma and brightness (as used on the GSW west facade), harmony between neighbouring colours in the spectrum and between opposing colours. The Natural Color System (NCS) literature suggests that harmony is built into the classification: 'Choose a series of colours with the same nuance from different hues. The colours will be perfectly balanced. This can be tried with other attributes, e.g. whiteness, blackness and chromaticness, but the results may not be as pleasing.' The navigation principles of the various colour models are discussed in Chapter 11, but many are underlaid by a desire to articulate the concept of harmony, ignorance of which has long been considered as the cause of errors in decoration.[25]

In 1828, David Hay, an active member of the 'Aesthetic Club of Edinburgh', published his book *The Laws of Harmonious Colouring*, which investigated the relationship between chromatic and musical harmonies:

> What are termed harmonious arrangements of colours, are such combinations as, by certain principles of our nature, produce on the sense by which they are perceived, the effect of pleasure or delight. In short, Harmony of Colours is to the eye what Harmony in Music is to the ear.[26]

This book, written by a self-described 'house painter', pre-dates Michel Eugène Chevreul's seminal text,[27] yet it contains many similar observations and suggests that there really should be both an understanding and an observance of the 'Laws' of colour harmony. In line with Chevreul, Hay is concerned with making the advice readily applicable to his audience. David Hay's suggestion was that the seven colours of the spectrum could be compared to the seven notes of a musical scale, with each colour being capable of forming a 'key'. Here, Hay is making reference to the seven colours of the spectrum as defined by Issac Newton.[28] Goethe's *Theory of Colours* (1810) suggests borrowing a 'tone' or 'tune' from music to set a colour key. The tone of colour in a painting, he noted, can be likened to a sharp key in music or to a flat key where a painting has a soft homogenous palette. Harmony, according to David Hay, was to be achieved by the use of opposing colours, while a melody would be composed of adjacent colours. He did not equate this to regularity, however, and here he coincides with Robert Venturi's observation that 'an artful discord gives vitality to architecture'.[29]

The link between colour and musical harmonies is a regular theme in discussions of colour.[30] Musicians will readily talk of the 'colour' of a passage, meaning the mood or expression. In 1911, Wassily Kandinsky[31] observed that 'the time for simple harmonies had passed', that the 'dirty dissonances of Schönberg [composer] would be tomorrow's harmonies'. Combinations considered initially as unharmonious or unacceptable may therefore be assimilated through education or exposure. Kandinsky tended towards an inclusive synaesthetic interpretation of colour, in which primary colours were assigned both primary shape and sound.[32] In his paper 'Uncertain Harmonies', Stephen Whittle describes the paintings of contemporary musician and artist, Kevin Laycock:

> he [Laycock] has consistently produced work in series, exploring a range of approaches through which the principles of musical composition can be adapted to the creation of abstract paintings ... While the underlying structure of his paintings is carefully, even scientifically, arrived at however it is then over-written with a wide variety of marks, touches and dribbles of paint. There is a constant interplay between order and intuition,[33]

which Whittle suggests is the equivalent of musicality.

1024 Farben, Gerard Richter (1973)

This technique has parallels with the grids of Gerard Richter and with the design of the irregular facades noted earlier. The grid, or series of regulating lines, provides a rhythm, but within which the final composition of the facades is intuitive and tends to be the product of an iterative exploration by the designer. Rafael Moneo's reference to music in his elevations for the Town Hall in Murcia, and Le Corbusier's well-known exploration of musical rhythms in his buildings of the 1950s, such as at the monastery at La Tourette (1959), can be seen as part of a continued exploration of these relationships. Le Corbusier's 'Modulor' was developed to relate the arbitrary metric system[34] to the human body, to bring what he considered to be the harmony and order of nature into play. As with Hay, Le Corbusier was providing practical advice based on a systemization of otherwise infinite choice, but he was aware that its application was a skill. He noted 'the "Modulor" is a working tool, a precision instrument; a keyboard shall we say, a piano, a tuned piano. The piano has been tuned: it is up to you to play it well'.[35] Le Corbusier himself was known to occasionally adjust a dimension against the dictates of the 'Modulor' to make fine adjustments to the spaces created.[36]

In his book *Interaction of Color* (1963), Josef Albers notes that any discussion of colour systems invariably leads to discussions of harmony, but he dismisses the link to musical tone.[37] He notes specific problems in the comparison with musical tone – as in the lack of any sense of time with colour. Notes fade and recede, whereas colours exist in space and can be read in any direction and at any speed. Albers suggests that this lack of a time factor in the composition of colours could be why it is not possible to define colour compositions by a diagram unlike notations in dance and music. Goethe also notes the dangers of comparisons with music, or between colour and sound, considering them to be more like two rivers emanating from a single source but divergent.[38] Despite these reservations, there is, of course, a direct relationship between wavelength frequency and colour. Seen as simple radiation, red equates to 626–800 nanometres (nm), blue to 450–480 nm, yellow 560–590 nm, and so forth. Neil Harbisson, a musician who suffers from an acute colour blindness known as achromatopsia, has been fitted with a prosthetic device nicknamed an 'eyeborg' which translates light wavelengths into sound frequencies and now refers to himself as the first 'sonochromatic cyborg'. The camera allows him to hear colour and, having now worn the device for several years, he is able to calibrate in both directions, his brain gradually tuning colour to sound, but also music to colour.[39] For the moment, this literal tuning remains a singular experience.

The pulse and rhythm of colour intervals across a facade are factors in the visual composition, as are the quantity and intensity of one colour in relation to another. These characteristics have parallels in music and art. It has been suggested, for instance, that the artist Philipp Otto Runge, who had such an interest in colour, believed that he could apply the principles of fugal musical composition to his paintings.[40] As related in Chapter 5, Erich Wiesner, a contemporary artist working with architects, has also made reference to Bach's fugues as a means of establishing rhythms. In referring to William Ostwald's complex 'Laws' for colour harmonies, the French cubist painter Amédée Ozenfant, who collaborated with Le Corbusier, makes comparisons with musical harmonies, but is quick to note that the great musicians consistently ignored, though were not ignorant of, the concept of consonant (permissible) and dissonant (prohibited) chords.[41]

Returning to the earlier question of how we judge successful colour combinations therefore, the key appears to be rooted in the subtle difference between 'Rules' and 'Laws'. Rules, it seems, can be useful as part of a design method. Laws are more finite. Once defined, even if self-determined,

Dynamic equilibrium of contrasting colours at the Fire and Police Station, Berlin, Sauerbruch Hutton (2004)

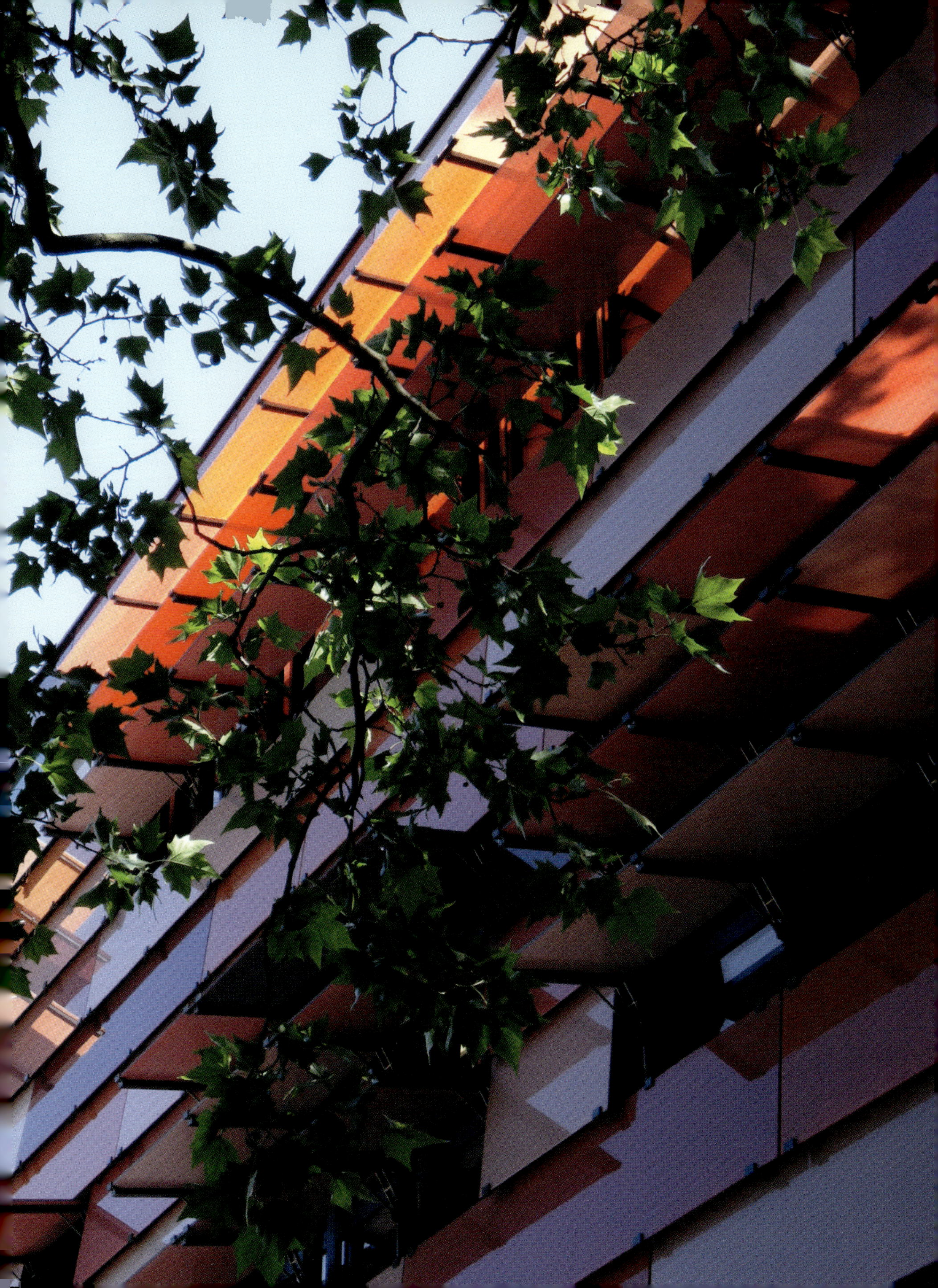

rules can be open to interpretation. 'By giving up preference for harmony, we accept dissonance to be as desirable as consonance.'[42]

Sauerbruch Hutton's Fire and Police Station in Berlin (1999–2004) uses two 'families' of colours based on the symbolic colours of the German fire brigade (red) and police (green). The building is an extension to an existing brick building. The new glass facade is made with overlapping tiles of back-painted glass. The material forms a sharp contrast in texture with the existing brick and stone building, and the angle of the shingle panels increases the reflectance from the sky. At night, the colour glows through the translucent section. The two colour families are wrapped from one side, which is predominately green hues, to the other, mainly reds and pinks. One colour gradually melds into the other using the irregular tile pattern. Occasionally, a dissonant tile is used to link one side to the other, but also, through simultaneous contrast, to heighten each colour by association with its opposite. The aim is to give their buildings

> a presence, which goes beyond the purely functional. And that is a juxtaposition which we find interesting ... it is more challenging ... to make a beautiful, articulate and sensual police station than it is to make an art gallery of that description.[43]

Detail of colour-coated steel panels at the Pavilion Heidi Weber, Zurich, Le Corbusier (1967)

As noted by Josef Albers, and echoed by Robert Venturi, dissonance can be as satisfactory as consonance. It can lead to a more dynamic and asymmetrical form of equilibrium, which gives a feeling of tension. Although Le Corbusier sought unity of form and colour, this was later clarified in the second edition of his *Salubra* palette in 1959: 'I find it useful to repeat this: nothing is more demoralizing than uniformity, a sign of stupidity. Nothing is stronger and more moving than unity'. In Le Corbusier's later work, such as the Salvation Army building in Paris, he makes more expressive use of materials and uses stronger colour, with black as well as white. He abandons his own rule of the wall surface having unity of colour in favour of colour applied to specific elements. He was seeking a designed equilibrium of colour, not necessarily through simple symmetries, but by using the intensity and appearance of certain colours to provide a visual balance. As noted by Arthur Rüegg, Le Corbusier was using colour in a more abstract way: 'Colour which is autonomous from wall or surface'.[44] This represented a notable shift in his use of colour from his earlier work.

Equilibrium

THE Museum for the Brandhorst Collection in Munich (2002–07), by Sauerbruch Hutton, also represents a shift in their use of colour. Instead of using one colour against another in panels, this complex facade uses long thin rods of a 40 mm square section, hung at equal distances and explores optical relationships of colour. The rods are dissociated from an undulating background cladding, with the result that each rod casts a serrated shadow. The visual effect of the rods, seen from a distance, is extraordinary. It is similar to the pointillist techniques of artists such as Seurat, where the colour mixes within the eye, rather than on the surface of the painting.

The eye seeks equilibrium in colour, just as it seeks equilibrium in composition and stability of structural forces. This may partly explain the effects of simultaneous and successive contrast, in that the eye is seeking to restore balance. Some optical effects that utilize afterimages, or other visual interactions between colours, are so stark that we find them astonishing, and there are many such examples circulating on the Internet. The most commonly known effects are the reversal of colours such as green to red, or purple to yellow, when observed intensely for a period and then seen against a neutral background. All colours, including black and white, have afterimages, but grey does not. Easily demonstrated with paint, the addition of contrasting pigments to each other will tend towards a neutral grey. Adding slightly more of one colour than the other will tint the grey towards that hue, but mixing opposing colours that are equal in brightness and intensity will result in a monochrome mid-grey. Mid-grey, therefore, can be seen as the equilibrium of all hues.

Josef Albers worked with his students to create a number of works to explore how contiguous colour is perceived by the eye. A difference between the actual surface and the surface as perceived. Using coloured paper, the students composed abstract designs, deliberately inviting ambiguity and

Reflected colours from the surrounding context at the Brandhorst Museum, Munich, Sauerbruch Hutton (2007)

Brandhorst Museum, Munich,
Sauerbruch Hutton (2007)

instability in the perception of colour. The intention was to explore the possibilities of using colour in composition and develop greater understanding in the students.

The exterior surfaces of the Brandhorst Museum meld into soft beigy colours from a distance. The shades, intensely different when seen close up, appear to mix together into a homogenous tone. Indeed, Hutton remarked that magazines have chosen not to publish certain views as the building tends to fade into a neutral sandy colour, merging with the surrounding buildings. Unlike their Sedus building, which has 20 colours randomly distributed across the facade, the Brandhorst Museum uses 'clouds' of different combinations of rods to try to produce areas of intensity. This is less about a rhythm of single colours on a surface than about the strengthening of the form of the building or of its various planes. Approaching close to the surface, however, the effect is much more activated and lively than a single coloured plane. Designing a gallery for contemporary art demands uninterrupted wall surfaces that can be dull and uninteresting, particularly in an inner-city area. Here, the three intersecting volumes of the cranked box form being differently composed do not read simply as planes, but break down the scale of overall volume.

The use of colour on the facades is also part of a narrative in relation to the art inside, which is mostly from the twentieth century and includes highly colourful studies by Andy Warhol, Sigmar Polke and a key collection of the work of Cy Twombly[45] that are hung in specifically designed rooms. Hutton relates the association:

> In twentieth-century painting, colour was being liberated from form, and colour had a value in itself as a spatial medium. In terms of spatial medium they're using not only the colours, different colours one against the other, but also using the background.

The background skin of their building is a perforated enameled sheet to improve the noise absorption. Its two tones then work with the colours of the rods in the foreground. The actual colours were chosen after extensive simulations and models at different scales and were surprising, even to the architects who had not used such 'strange pistachio milky colours and strange ice creamy pinks' before. The colours are not chosen individually, but purely in relation to the optical effect when placed contiguously and against the different backgrounds.

The sense that colour is unstable, temporal and, to some extent, unpredictable – in that we cannot know with absolute certainty how others perceive colour – adds to the complexity of its use. A strong colour next to a mid-grey will appear to tinge the grey. Placing two complementary colours adjacent to each other can provide balance, but each colour will appear to shift slightly towards its opposite, and this could dull the vibrancy of each. White set next to black can appear as grey; but one white set against another can also appear grey. As noted by Johannes Itten, 'their stability is disturbed – colour is … dematerialized'.[46]

The work of Sauerbruch Hutton is specifically characterized not just by its use of colour, but by the polychromatic combinations. They tend not to use colour in isolation, preferring the interaction of one colour on another. Although they have read widely in respect of colour theory, they are aware that no single theory drives the work. It is not dogmatic; rather the colour seems so wholly intrinsic that it is inseparable from the architecture.

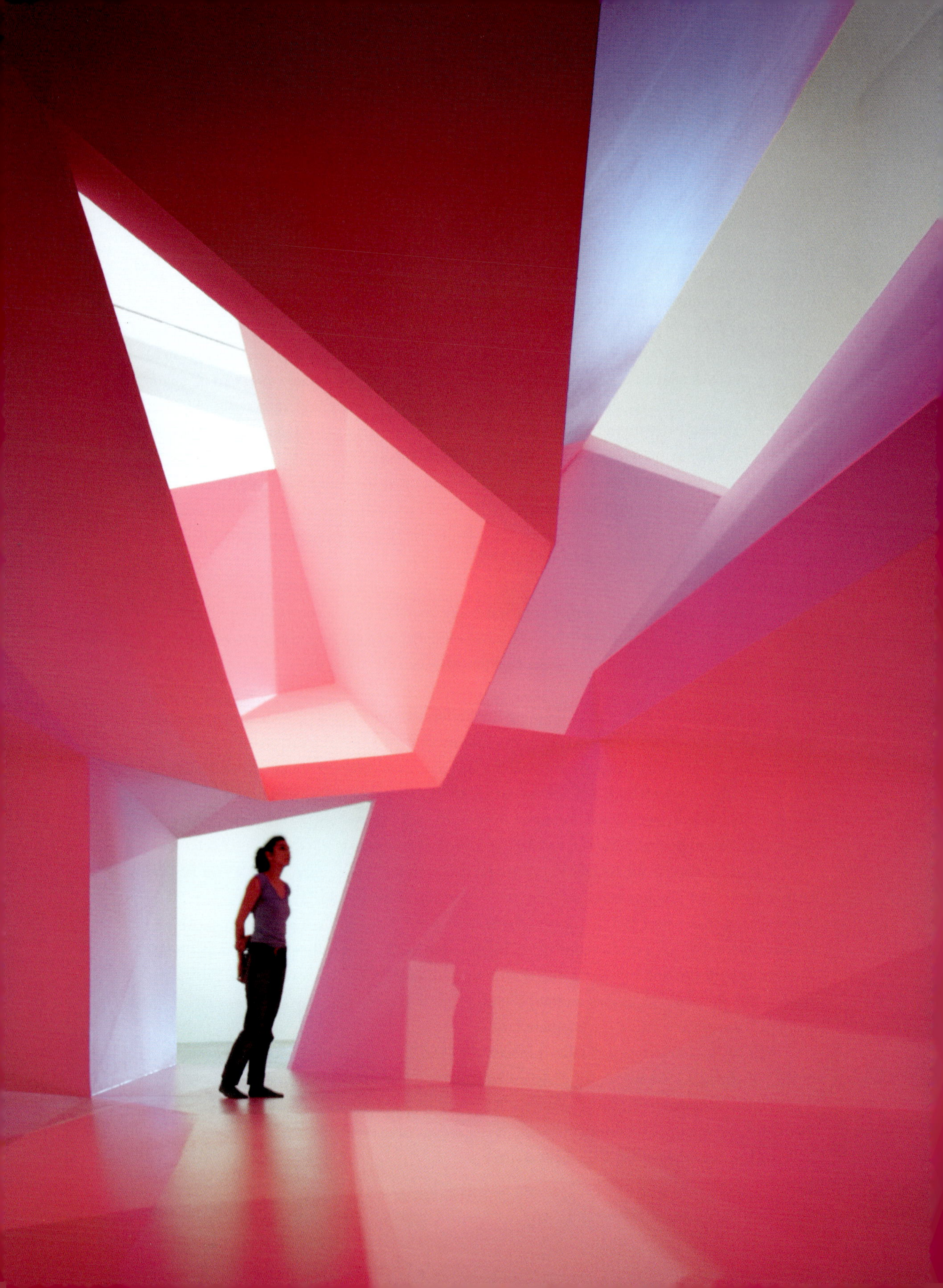

– 10 –
Transformational, instrumental colour: UN Studio

ARCHITECTURAL design education tends to focus on the creation of space, with surface treatment seen as of secondary importance. Where once the plan was seen as the generator of space, computer modeling now allows architects and engineers to design in three dimensions. Although representation of space is becoming easier, for many architects the choice of specific colour still comes late in the design process and its effects are not easily replicated. Colour, surface and texture as integral parts of our experience of space are used by the paint industry to present the creative potential of colour to alter mood and association in a manner that is intended to trigger memory and emotion. Yet, the potential ability of colour as an instrument to transform space and surface is often left unrealized by architects.

The work of UN Studio is an example of colour used in an instrumental and morphological sense, wrapping exteriors and immersing the building user internally, through the manipulation of colour and space. Their Agora Theatre in Lelystad (2007), north of Amsterdam in the Netherlands, provides evidence of an interest in colour that stems from Ben van Berkel's early training as a graphic artist. This chapter considers recent projects, such as in the new town of Almere, which uses an iridescent film applied to glass, their music theatre in Graz, with a colour palette based on theatrical make-up, and two department store buildings in Taiwan and in South Korea where computer-controlled lighting is used to transform the appearance of the buildings over time.

Amorphous space

MODERNIST architecture shifted the emphasis from the design and ordering of rooms to open-plan, framed spaces. Facades became free of their load-bearing function and space was immediately free to flow. Rather than defining the atmosphere or function of one room in terms of another, as would have been the traditional approach, and as noted previously in Chapters

facing page
Holiday home project, UN Studio, (2006)

2 and 5, Le Corbusier wrestled with how to use colour in spaces no longer defined simply by walls. His Villa La Roche, Paris (1923–24), demonstrates colour used as a rarifier, identifying specific architectural elements, such as a chimney or the curved rising ramp. These elements move through space in three dimensions and the colour aids the legibility of the composition.

More recently, computer modelling, structural analysis and computer-controlled manufacturing have both enabled and spurred a growth in formal manipulation. Some of the products of such form-making games are criticized as crude and without context by architects such as Caruso St John. At the same time, the ability to construct space that flows not just vertically, but in every direction, sometimes uninterrupted by conventional vertical structure, has produced some extraordinarily dynamic spatial experiences.

Patrik Schumacher, partner to Zaha Hadid, refers to their work as 'articulated complexity',[1] but argues that the spaces are highly ordered. Indeed, he suggests that their practice is extending the radical openness that was the key liberation of the Modernists and de Stijl. Hadid's early paintings drew on Constructivists such as Malevich and El Lissitzky. The slow, smooth curves one associates with Hadid are inspired by structures of the Russian sculptor Noam Gabo (1890–1977), as well as by digital abstractions of natural landscapes. In Hadid and Schumacher's work, applied colour is used to emphasize specific fluid forms, as in the black, sinuous, curving roof beams at the MAXXI Museum of Twenty-First Century Art in Rome (1999–2009), or the black, shifting cuboid and stair elements at the Cincinnati Contemporary Arts Center (2004).[2] In most cases, however, they prefer to use bare materials, giving preference to form over surface.

MAXXI Museum of Twenty-First Century Art, Rome, Zaha Hadid Architects (2009)

UN Studio,[3] a multi-disciplinary practice based in Amsterdam, has experimented with continuous surfaces, where wall becomes floor becomes ceiling in a number of projects. The small pavilion for the Venice Biennale in 2008, the Changing Room, explores the transformative possibilities of material. An earlier project, the Mobius House (1993–98), is a sinuous band of accommodation twisted along a linear path. Public and private spaces are given unexpected relationships to each other in section. The Villa NM in New York State (2006), destroyed by fire in 2008, twisted a horizontal floor plane to vertical, like a propeller blade. UN Studio's most extreme project to date, the Mercedes Benz building in Stuttgart (2001–06), wraps concrete surfaces around a rising helix. Conventional sections and plans, even photographs, are unable to convey the complexity of the spatial configuration. The architecture is disconcertingly unreliable:

> where depth is suggested it tends to be in the surface treatment rather than an expression of inside on outside or transparency to that behind. ... throughout the project, treatments of shape and form reinforce suspicions that the architecture is reliable neither in expressing the truth of its own constitution, nor in producing, containing or directing human activities, practices and encounters.[4]

If Le Corbusier suggested that the free plan required a reconceptualization of colour, what do these new complex spatial experiences mean for colour?

Surfaces are not flat, but continuously curving or faceted and, in such circumstances, colour is even more unreliable in relation to our visual perception. The flowing surfaces mean that colour is not seen in blocks or consistently coloured planes; rather its appearance shifts as fluidly as the form. Although people are now experiencing a type of space that has not, until recently, been possible to construct, human physiology remains the same. The space is more immersive and dynamic, and demands that the viewer moves through space in order to fully appreciate the complexity. These are restless forms, and colour becomes even more ambiguous if used in such spaces. Perhaps we should recognize that this is the case and explore the effects that might be created. UN Studio has begun to experiment with applied colour in a number of recent projects and offers insights into how colour can be used as an instrument to transform our interpretation and experience of three-dimensionally fluid space.

According to Ben van Berkel, the renewed interest in colour did not spring from the colour itself, rather it was as a means of exploring ideas of space and time. For Sigfried Giedion in the Introduction to his seminal book *Space, Time and Architecture* (1941):

> it is the interaction between volumes which gives full orchestration to the first architectural space conception ... volumes affect space just as an enclosure gives space to an interior space. ... that architecture is approaching sculpture and sculpture is approaching architecture is no deviation from the development of contemporary architecture. One of the features of this evolving tradition is the simultaneity of freedom and involvement.

While Giedion was referring, at the time, primarily to high Modernism, his text can equally be applied to a further evolutionary shift evidenced in recent architecture.

Common to most of UN Studio's work is a sense of movement, whether essential to the project, such as infrastructure projects like the Erasmus Bridge or Arnhem Central Station, or implied in the twisting geometry of the smaller pavilions and houses. Space, therefore, is rarely static, it changes in time and motion:

> We are fascinated by the possibility that anything can be translated into architecture … If you establish effective translation techniques you can translate anything. To do this effectively you need an instrument akin to a kaleidoscope, that will twist the thing beyond recognition. This is the place where real discoveries are made.[5]

It is entirely fitting that van Berkel and Bos have begun to develop a passionate interest in colour. The concept of an inclusive practice lies, not just in organizational terms, but in the way that they draw upon a very wide range of sources from everyday life. Art, sculpture, culture, music, shopping and philosophy provide them with opportunities to exploit each field for what it will yield, and for what it can contribute to their architecture, before moving on to another area of study.[6] They seem comfortable with, or even to thrive on, a continual process of renewal and transition. Colour, therefore, is one more in a series of possible instruments that can be translated and employed in an integrated way as an agent for architecture.

The instrumentality of colour

BEFORE considering the use of colour in the complex geometries of UN Studio, it is perhaps useful to remember the radical effect of colour on the quality and perception of more conventional architectural space.

Using colour internally to 'tune' a space to a particular atmosphere is an interest of the Irish architect Grainne Hassett. Brookfield Community Centre in Tallaght, a socially deprived area to the south-west of Dublin, is an example. Rather than seeing the use of colour as defining, clarifying or coding elements of form, Hassett disrupts the unity of architectural elements, and the legibility of the space, through the interaction of adjacent colours. Part of the intent is to alter mood and to invigorate and stimulate the users. Hassett is aware that her interest in colour has emerged only recently, but might always have been there, albeit subliminally. She has always had a strong sense of tuning up the emotion, or the sense of the place in her architecture.[7]

Hassett has undertaken research into colour as part of an Irish Arts Council project for a recent publication, and has explored the work of Donald Judd, Bridget Riley, Kazuyo Sejima and Mies van der Rohe. For Hassett, it is Judd's use of colour to dispel readings of mass, and readings of weight and, sometimes, readings of gravity that is highly motivating. The making of objects that appear simultaneously unstable yet incredibly stable is, for her, a source of pleasure and intrigue.

Brookfield Community Centre, Tallaght, Dublin,
Hassett Ducatez (2008)

She is beginning to explore the idea that colour can distort or nuance the optical qualities of a space, rather than make it legible.

In Hassett's case, the context of the Brookfield project, both physical and social, was justification for the intensity of the use of colour. The building is set in a bland and impoverished context, and is predominately used by young people. There was, therefore, a profoundly social agenda to the project. The building is intended to be a focus – stimulating, invigorating and with a strong identity. The project was selected for inclusion in a film, *The Lives of Spaces*, as part of the Irish submission for the Venice Biennale in (2008). As noted in the submission, the film 'makes evident architecture's great central responsibility – the shaping of the spaces that in turn shape society' and responded to the theme of the Biennale 'Architecture Beyond Building' by asserting that architecture extends beyond building to both 'embrace and engender life'.[8]

Brookfield Community Centre is a very low-budget building, and colour became a way in which the otherwise simple spatial characteristics could be intensified, and a clear social identity of an inclusive institution established. Hassett was conscious that she did not want the paint to appear cheap and, although colour choice was restricted by budget to standard colours, the atmosphere is 'tuned' using unlikely combinations. Hassett's research work included artists and architects whose work she found more directly related to practice than that of colour theorists. Like fellow Dubliners O'Donnell and Tuomey, she has observed that one of the difficulties architects have in working with colour is the impossibility of representing colour to scale. Light cannot be scaled down in a model and, as colour is effectively light, it is impossible to represent the full effect of colour on space. Donald Judd, she notes, went straight to full scale. He had to work from his mind, and he had to work from his register of colours learnt as a painter and continually being observed. I do think this is a key issue because, as architects, we don't train the visual mind.[9]

Hassett had the advantage that the project was slow to construct, so the colours could be chosen on site. In conventional construction projects it is rare for clients to fund a process of trial and error. This may partly explain why many architects approach colour so nervously.

Experimentation

BEN VAN BERKEL AND CAROLINE BOS set up their office in 1988 after meeting during their training in London. Bos had studied History of Art and offers a rational, critical perspective complementary to Ben van Berkel, who acknowledges that he is more intuitive by nature. His knowledge of colour theory stems from his early career as a graphic artist. One of his tutors had studied for a year with Johannes Itten at the Bauhaus, and this early training gave him an enduring, although for some time latent, interest in the possibilities of colour. He also has developed a habit of drawing, without necessarily knowing where the drawings will lead, more as a process of self-contemplation. The results are not necessarily sketches of anything, but a way of allowing ideas to surface in a similar method to that used by Will Alsop. A genealogy might be traced to the automatic writing of the Surrealists and similarities, not in output but in method, to the drawings of Coop Himmelblau.[10]

Bart Lootsma has described the work of UN Studio as 'a unique mixture of conceptualisation and expression, of rational considerations and intuitive decisions'.[11] Accordingly, UN Studio's architecture moves in cycles, and they appear very open to discussion, to conversations that might steer them in a new direction. They are aware also that their work is criticized for an apparent lack of coherence. It moves, shifts and adapts. Early projects showed no sign of any interest in colour. Moving from graphic design to architecture, Ben van Berkel soon learned to keep quiet about colour. His training had taught him that form, space and materials were paramount, that space was defined and manipulated by form and light, and that surface was merely a resultant of form. Surface treatment was seen in most architectural training at that time as secondary or even tertiary to space and material. Embarking on architectural practice, spatial organizational qualities were therefore of primary interest. Colour came later, although when it did emerge, it came quite boldly, 'having learnt to control all the ingredients, we have radicalised some of the aspects much more'.[12]

Only recently has this interest in colour been allowed to develop as part of a continual process of renewal and reflection. Only now, when they feel that they have mastered the basics of architecture, can they contemplate the addition of a further layer of complexity. Colour is used to influence the spatial tensions within a project, or the appearance of space and form. They also acknowledge, however, that, as an architect, one may simply not entirely understand what one does and why one does it. Critical practice comes by experimentation or action, followed by a period of reflection. We see this in the work of the artist Bridget Riley, and also in the architecture of UN Studio. To imply that there is not necessarily a reasoned argument, and that they are prepared to work intuitively, suggests that they place a high value on the search for originality and creativity. At the same time, the office is comfortable with a degree of post-rationalization, seeking to learn as they experiment. The educationalist Donald Schön analysed the demands of a representative professional practice and showed that, in reality, most problems are 'by nature, messy, ill-defined, uncertain and invariably unique; and that solutions to these problems call on the integration and use of knowledge from many different domains'.[13] Schön also questioned the difference between 'professional knowing' and academic knowledge as presented in journals and textbooks, as a means of understanding the gulf that practitioners feel between academia and practice. Central to Schön's argument is the proposition that professional knowing is maintained and enhanced by a process of 'reflection in action'.

It was clear to van Berkel and Bos early on in their practice that the traditional fragmented organizational structures of architectural practice and the construction industry would not support their aims. They were too restrictive and not conducive to experimentation. After working on the Erasmus Bridge project, they transformed the nature and name of the practice to reflect a more contemporary, inclusive and integrated design philosophy. Through communication and digital networks, international teams of consultants and their computers are linked in order to integrate structure, light, geometry, materials and paths of movement and so challenge conventional practice.[14] This structure has allowed them to take on very large infrastructure projects, such as the Arnhem Station redevelopment, but also appears to have given more freedom to explore the architectural content, including colour. One of their most extraordinary projects is in Lelystad.

Illusion

THE town of Lelystad, north of Amsterdam, has a very short history. Constructed on a reclaimed polder in the 1960s, it is surrounded by an extraordinary landscape of flat, featureless land with very little vegetation. It has not yet reached its target population and suffers from this low density of population in comparison to the very high intensity of occupation common in the Netherlands. A recent master plan, by landscape architects West 8, aims to establish the town as a more attractive prospect for living and working. Part of this transformation involves a denser centre with irregular street configurations leading pedestrians from the station through a network of narrow streets, much more in keeping with a traditional Dutch town. This change is already taking shape and includes the development of a new theatre designed by UN Studio.

The 1960s town was almost entirely muddy brown in colour and Ben van Berkel's initial feeling was that it lacked any sense of place. According to the architect, the commute between Lelystad and Amsterdam provided part of the answer. In this landscape, the sky is dominant. It has

Agora Theatre, Lelystad, Netherlands, UN Studio (2002)

facing page
Main theatre interior

been painted by artists for centuries and the quality of light over the Isjelmeer (the large inland sea gradually being turned into land) is quite exceptional. Given the quality of the sky, particularly the yellow and orange glow at sunset looking west over the expanse of the North Sea, which Ben van Berkel saw as 'beautiful, bright, wonderful and optimistic', he could not understand why building colours should relate to the clay of the earth rather than the colours of the light and sky.

The project came at a transition point, or was perhaps the instrument through which the practice unlocked a previously dormant interest in colour. Until then, the architecture was predominantly about form and natural materials. The Agora Theatre is a striking orange, amorphous building. It could not be more alien to the earth-brown town. The intense colour has divided the population. Those who had the courage to commission and use the theatre are fiercely supportive and proud of this unique, folded structure that, without a doubt, gives the town cultural credence as well as directly challenging the monotony of this previously sleepy satellite. Those against are utterly

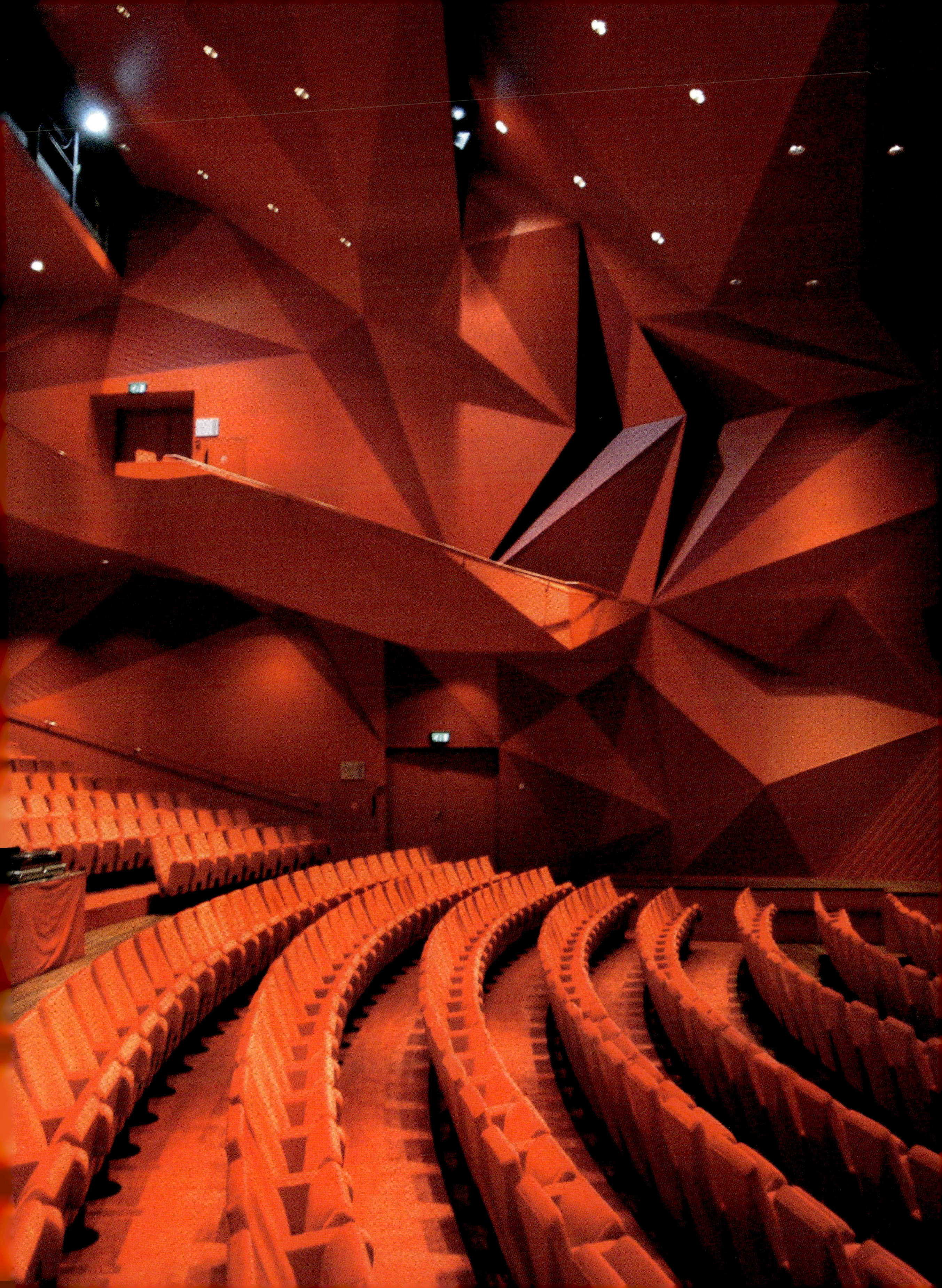

bemused that someone has been permitted to build a giant orange blob in their town centre. Colour can, of course, provoke intense controversy, and it is rare to see such a bold intervention. The town that was once bland, brown and unable to attract people to live there now has an evolving landscape and a cultural building that is an unmissable symbol of change.

The colour is particularly striking, given both the scale of the building and the folded plate form. Colour is used to further intensify the effect by employing nine shades of orange on the facets. Although at certain times of the day it may appear to be all one colour, with the effect of daylight appearing to create different shades, this is an illusion. The colours are actually tonally different. In this case, the surface folds and is modified to intensify the reading of form. As if that were not sufficient in making the theatrical event, further illusions await internally. The outside skin and inside space are unrelated. The main theatre, secondary theatre and foyer form a series of independent volumes, loosely related within the overarching skin. Architecturally, the spaces are ambiguous. There is no attempt to be honest in expressing volumes and activities inside on the outside. Even the flytower is disguised, almost entirely, by the orange mask. All is not what it seems. It is complex and contradictory. The theatre also seems to face the wrong way, turning its back on a small square and, in doing so, perhaps failing to engage in the most obvious social space. Perhaps, as the master plan develops, the orientation will be explained.

The colour of the theatre is part of an intellectual conversation about the nature of theatre, as opposed to an end in itself. Coming from a musical family, Ben van Berkel had spent a childhood in and out of theatres, an experience that appears to have left a strong impression on him. Theatre is something very special to him, and a visit to a theatre an event that should be celebratory and memorable. In each volume of the Agora, a different colour is dominant. The main theatre is draped in an intense lipstick red that is enveloping and immersive of the audience. Red is a plush, deeply traditional colour for a theatre, but here it is a much clearer, sharper tone. As with the external facets, the panels appear to have been folded into shape and are again subtlety coloured in different shades of red to suggest greater depth. The viewer is deluded into assuming it is one red, when the colour is used as a carefully choreographed surface treatment that intensifies the three-dimensionality. The colour itself could be seen as an 'intellectual illusion'.[15] The secondary theatre uses a clear primary yellow and black. One immediately understands that this is a more experimental space, where audience and players may merge. It is simple and rectangular in form, allowing a number of different configurations or functions.

Interrogation of the meaning of contemporary theatre is also apparent in UN Studio's Music Theatre in Graz, Austria (2008). The pigments used in the interior spaces were derived from the luscious purply-browns, reds and whites of make-up. Van Berkel notes that the colours are deliberately chosen to 'bite' each other and makes reference to the Japanese use of make-up as a mask, highlighting the face within darkened spaces. The interior of the theatre is wrapped in brown, the staircase a sinuous red, while the exterior mesh is illuminated by changing light colours of a less subtle range.

facing page
Lipstick-red stair and purply-brown interior,
Music Theatre, Graz, Austria,
UN Studio (2008)

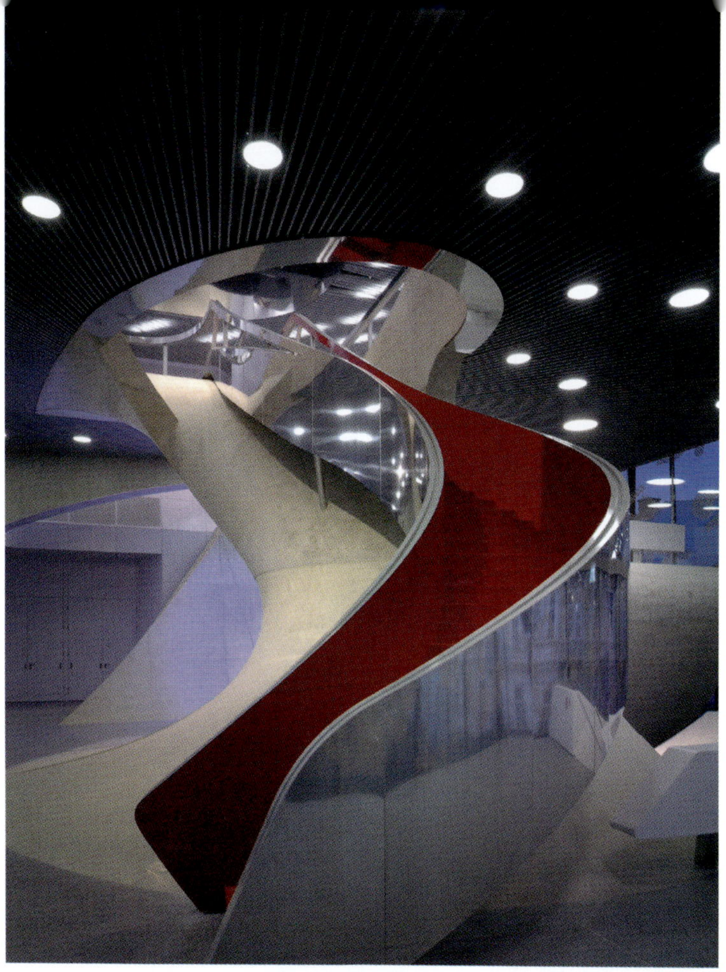

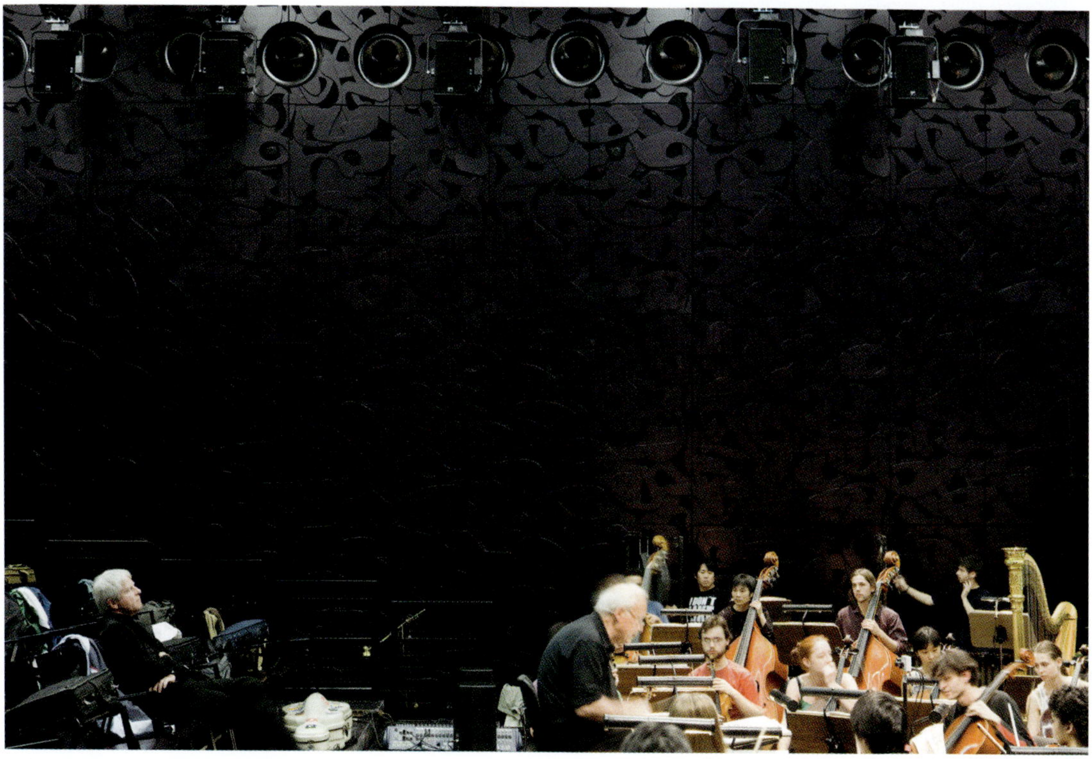

At the Agora Theatre, the foyer snakes vertically, connecting the various volumes and culminating in a large skylight. The outside face of the main stair is painted in a bluish pink. Yet more deception. The pink appears to be gradated in tone from an orange-pink at the ground floor to a strong purple-pink at the highest point. It is all one shade of pink. Only the changing light from artificial to natural affects the perception of the colour. Caroline Bos recalls the difficulty of persuading the contractors to apply the pink paint. By then they were proud of their work and were convinced this would diminish the quality of the space. In practice, this pink ribbon seems to accentuate the verticality of the space and leads the visitor through this nebulous assemblage. The shifts in chromatic value have an intense effect on the space, and on the perception of the space through which the pink band moves.

The theatre foyer is recognizable as part of a similar dynamic composition to the Mercedes Benz Museum (2001–06) but in all other respects the building represents a shift in their work. The practice had previously negotiated a period they refer to as their 'blue period', which included projects such as the Mobius House (1998), Het Valkhof Museum (1998) and the Erasmus Bridge (2003). The pale blue of the bridge, now symbolic of the regeneration of the south of Rotterdam, reacts with a chameleon-like effect to changing light conditions, to appear either intensely white, pale grey or deep purply blue. UN Studio are acutely interested in this transformational effect of colour. Far from being concerned or nervous about its unpredictability and unreliability, they use colour precisely because of its ambiguity. The bridge also appears to transmute in relation to other variables, such as the distance and angle of the viewer and the water, wind and daylight conditions. The 'baby blue' colour of the bridge was chosen partly because of its ability to express such apparent chromatic change.[16]

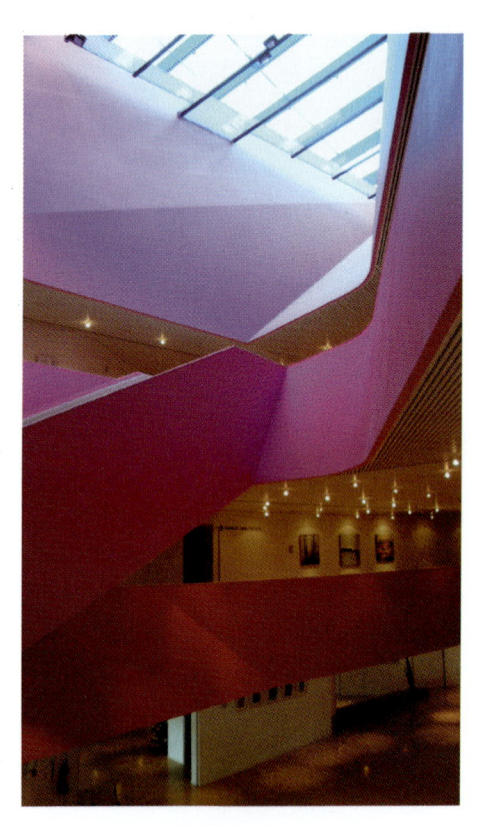

Pink-painted stair in the foyer of the Agora Theatre, Lelystad, Netherlands, UN Studio (2002)

Transformational colour

UN STUDIO'S experimental approach to architecture and to colour is further exemplified in a project for an urban block of office accommodation at Almere, another new town north of Amsterdam. Almere is one of the youngest cities in the Netherlands, established between 1976 and the mid-1980s. Rem Koolhaas/OMA prepared a further masterplan for the town in 1994, which is now largely built and which plays on the original intention of a multi-centred town. To the north of the station is one large expanse of development; the second, to the south along the waterfront.

3M Radiant Colour/Light Film used at La Defense
office development, Almere, Netherlands,
UN Studio (2004)

Despite these large-scale high-density interventions, Almere still has a strange suburban *Truman Show* quality.[17] Surfaces are clean, regularly scrubbed and lacking in most normal signs of humanity. There is an eerie architectural zoo of quirky new building emerging from a tilted ground plane. Children on skateboards enjoy the one, albeit artificial, slope in an otherwise flat landscape. The highest building is by Gigon/Guyer with artist Adrian Schiess and, as with their other projects with this artist, it is highly colourful and already an established icon for the OMA-planned downtown area.

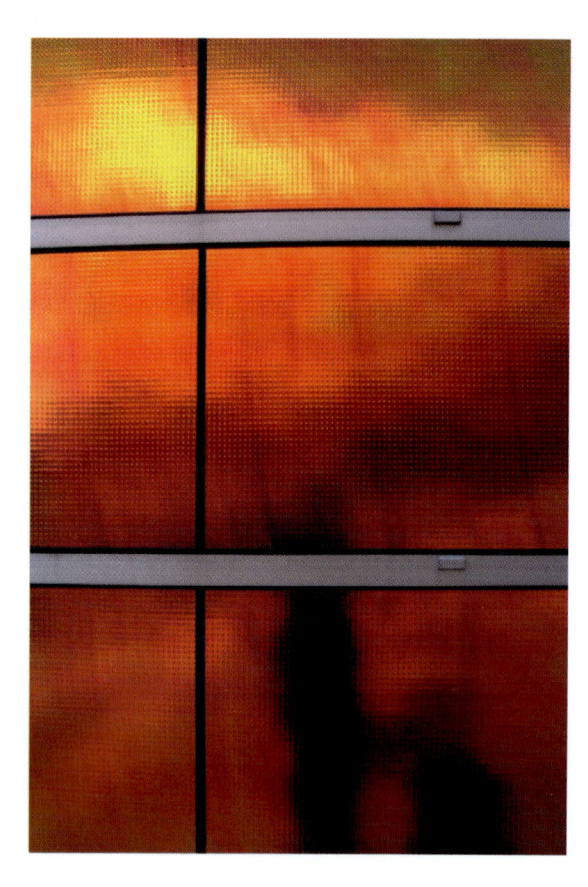

Office development, Almere, Netherlands. Although all one material, the panels appear to change colour depending on the angle of the viewer and the light conditions from deep red to orange, green and yellow, UN Studio (2004)

To the north of the station, a mix of commercial and housing developments is nearly complete, increasing the density. UN Studio's project sits within this neighbourhood. Shiny grey aluminum on the exterior surfaces of the block is slashed with oblique cuts across the centre to reveal a highly colourful series of snaking void spaces. The building appears inside out, bland and homogenous on the outside, vibrant and astonishingly dynamic on the inner faces. Most entrances are also on the inside of the urban block, forcing the visitor to circumnavigate the exterior. The inner skin is a pure experiment. They recount that when visiting 3M's laboratories in search of possible materials, van Berkel and Bos spotted reject rolls of a test material that had not yet found a use. Now known as '3M Radiant Colour/Light Film', it was originally intended for wrapping perfume bottles. The film, in this case lining a glass facade, reflects light in such a way that the colour of the material appears to change as the viewpoint and light change. Viewed perpendicular to the skin, it is a clear, deep red, sideways a green, yellow or blue. The reflected light from the panels is itself so intense that it diffuses across the pavements and internal courtyard spaces, infecting the grey surfaces.

facing page
Projected coloured light on the facade of the Star Palace,
Kaohsiung, Taiwan, UN Studio (2008)

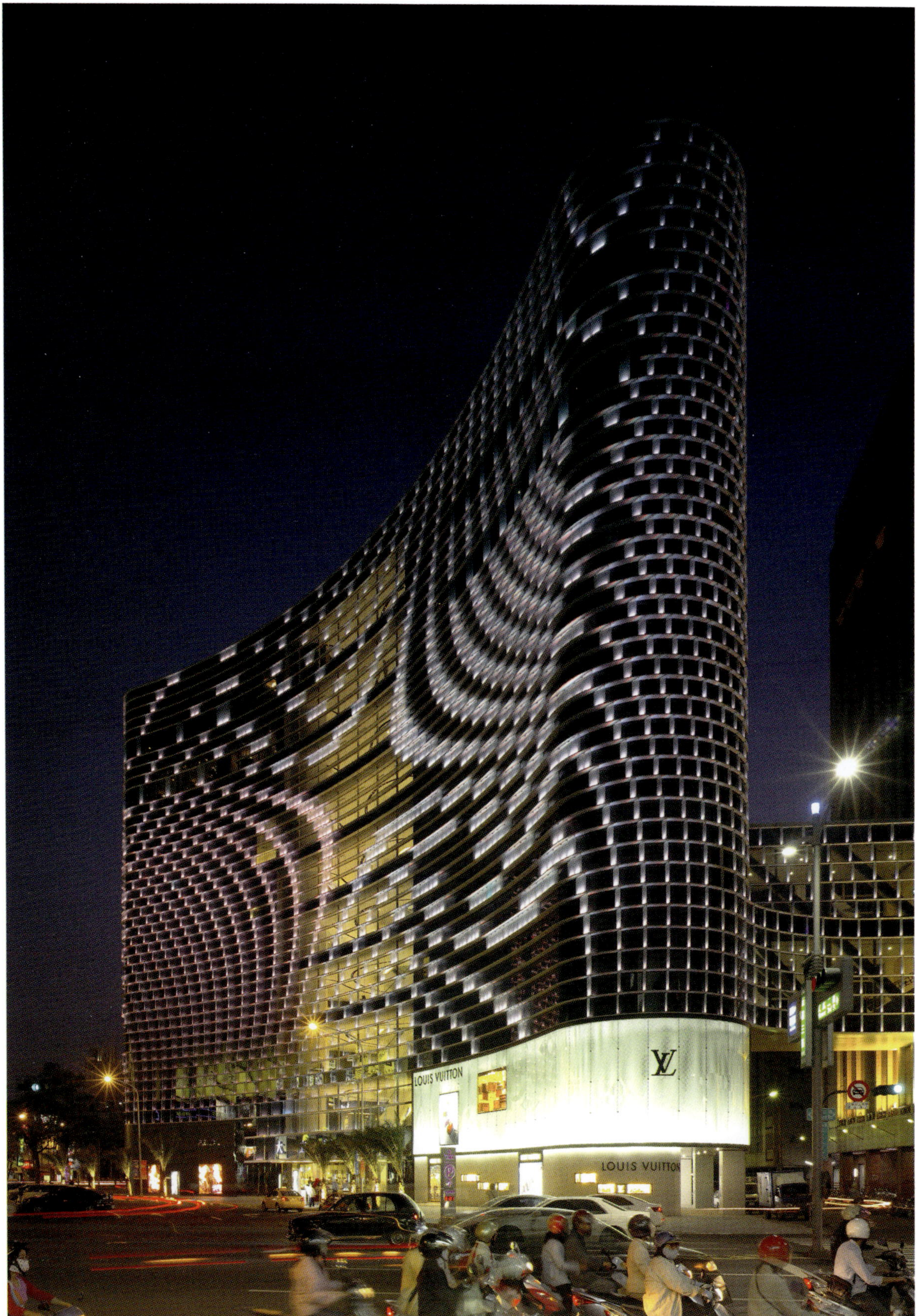
LOUIS VUITTON
LV
LOUIS VUITTON

Without the material, and the resultant colour, one could argue that this is a fairly soulless perimeter office building. The courts are barren, without any form of softening landscaping. It is a disconcerting experience to dwell in the voids, feeling constantly observed by acres of mirror sunglasses. The coloured light bounces around the faces of the courtyard and into the edges of the rooms inside the buildings, which have now been occupied by social security and tax offices. The light, without doubt, modifies the feeling of the spaces and may serve to lessen the potentially humiliating experience of claiming social benefits. The colour on this building completely transforms its appearance, and the appearance constantly changes with the clouds, the light and the position of the viewer.[18]

There has been a recent growth in the use of coloured light to transform architecture externally, spurred on by the ability to programme complex lighting combinations digitally. UN Studio has completed two department stores, the Galleria in Seoul, South Korea (2004) and Star Place, Kaohsiung Taiwan (2008). In both, projected light is used directly to distort and modify the appearance of the building by day and by night through timed pulses. Ben van Berkel relates the changing colour to the seasonal pulse of fashion in clothes, using the same metaphor of dressing for a proposed apartment building wrapped in ribbons of steel in New York. In the case of the department stores, however, the pulse is much faster than any seasonal variation in clothes. Through the constantly changing light, the authority of the building image is placed in some doubt. Who is in control of the image: architect or computer programmer?

Van Berkel admits that they had not entirely understood how easy it would be for the client to change the appearance and so, for the second building, the practice constrained the lighting design through a tighter contractual agreement, effectively legislating the colours and thereby ensuring architectural authorship.[19] There are, clearly, inherent dangers for architecture in this drift towards brand and image over substance and materiality. Taken to its limits, architecture becomes mere lifeless surface by day and, by night, a 'brandscape' for projected colour and light.[20] It may still be experiential, but only in a cinematic sense, and is easily open to manipulation. These technologies introduce opportunities, but also bring new dilemmas for architects more accustomed to permanence, durability and stability. By contrast, the joy of the reflected colour at La Defense in Almere is that it is generated entirely by daylight. Although concentrated, it is more subtle, less controlled, and not programmed but utterly contingent on the vagaries of natural light.

UN Studio's spa hotel in Castel Zuoz at Engadin (2004), near St Moritz in Switzerland, combines a coolly modern apartment building, a series of gliding horizontal planes, pale-green panels on balconies and an intense corporeal experience of the Hammam spa inside the original hotel building. The colour on the exterior was designed to be seen against a dynamic context: silver-white snow or the fresh green landscape that emerges each spring. For the spa, Ben van Berkel made watercolour drawings of the design, dripping one colour over another. When full of people, van Berkel suggests that it is like being inside a painting. It has a very intimate and intense quality and the colour of the surfaces, modified further by coloured light, bathes the bodies in a very sensual wash. Like the interior of the Agora Theatre, or within a James Turrell installation, the combination of people and intensely coloured space is highly charged and extremely memorable.

Gilles Deleuze considered this effect to be entirely the role of space.[21] Space is more than fixed geometry, more than simple built form. Architecture is multi-dimensional; although stable, it is experienced in a transient way and should allow multiple subjective readings and experiences. Unlike some of the other practices interviewed for this book, UN Studio appear to be entirely open to the evolution of the work of the practice in a series of leaps, periods of intensity and exploration of themes. There is less consistency, interests come and go. Colour has recently become one such experiment and their interest is rooted in its instrumental capacity, in what one can do with colour to transform architecture, to paint space, to fill each colour with architectural, theoretical meanings. In these examples, colour transforms space and surface. Their architecture is moving, rather than static, constantly changing, contingent, dematerializing and reconstituting.

Hammam spa at Castel Zuoz, Engadin, near St Moritz, Switzerland, UN Studio, (2004)

facing page
Hotel exterior at Castel Zuoz,
Engadin, near St Moritz,
Switzerland, UN Studio (2004)

Farbenkugel.

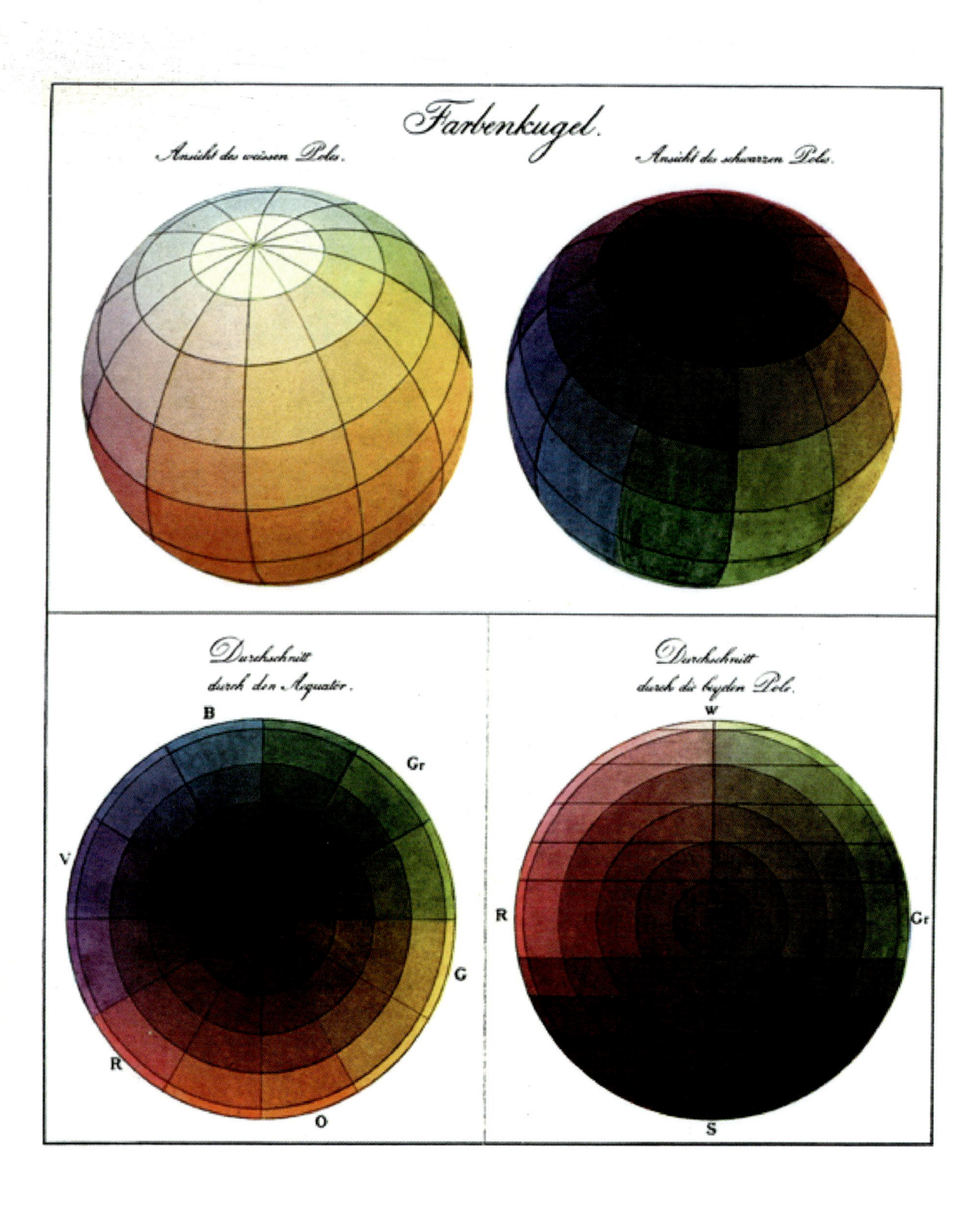

Ansicht des weissen Pols. *Ansicht des schwarzen Pols.*

Durchschnitt durch den Aequator. *Durchschnitt durch die beyden Pole.*

– 11 –

Navigation, communication and language

GIVEN the wide range of paints and pigmented materials available, architects and designers must learn to navigate around and between the multiple colour systems and terminologies that have been generated to describe colour. Each paint manufacturer, for example, has a different method of indexing and cataloguing colour, despite numerous attempts to standardize. Manufacturers can copyright a colour by name, but not the colour itself, and so many continue the practice of having not only a name, but also an internal or technical code. The nature of most construction contracts, where an element of choice lies with the contractor in the selection of a supplier, means that the architect or designer may not be able to define colours precisely until a very late stage. Commonly, choices are made in relation to building sequence or component delivery times, meaning that the selection of structural systems and construction elements inevitably comes before the consideration of surfaces finishes. This frustrating, practical problem contributes to the lack of commitment to colour at the early stages of most projects. Orientating oneself within the colour maze is far from simple, and it is useful to be aware of some background to the most common classification systems and the general language of colour.

Navigation systems

IN 1803, on a visit to Weimar, Philipp Otto Runge, a young artist, unexpectedly met Johann Wolfgang von Goethe and the two formed a friendship based on their common interests in art and colour. In his text *Farbenkugel*, published in Hamburg in 1810 (the same year as his untimely death at the age of 33), Runge defined and illustrated a faceted colour sphere. It was one of a number of attempts by a succession of artists, chemists, physicists and physiologists to articulate colour in relation to mathematical and geometrical relationships. Perhaps due to his early death, Runge's text is less well-known than Goethe's *Theory of Colours*, published later the same year, although each

facing page
Farbenkugel, Philipp Otto Runge (1810)

acknowledges the other within their respective texts. Goethe's principles have been dissected and criticized – as he himself had disparaged those of Sir Isaac Newton – but broadly the same geometrical models of colour representation are used by the professional today.

Before Newton, colour science was largely still attributable to Aristotle.[1] Unlike Newton, Goethe considered colour to be a visual phenomenon, happening in the eye and mind, rather than as an empirical aspect of light.[2] Where Newton sought origin, Goethe sought meaning.[3] His work focused on physiological colour, on what the eye sees, on visually opposite colours and the effect of afterimages. Goethe was severely criticized at the time for daring to question Newton, who was highly respected and whose theories had been gradually incorporated into the Royal Academy of Art's curriculum.[4] Among Goethe's stinging criticisms was the charge that Newton alluded to only seven colours, rather than a continuous spectrum, perhaps being too ready to make comparisons with harmony and musical scales. Newton, he suggested, conveniently and arbitrarily added indigo to the six more obvious hues. Goethe suggests that a practical man, such as a dye maker

> who experiences benefit or detriment from the application of his convictions, to whom loss of time and money is not indifferent … such a person feels the unsoundness and erroneousness of a theory much sooner than the man of letters.

He continues in a more direct criticism of Newton:

> this was an additional evil. A great mathematician was possessed with an entirely false notion on the physical origin of colours; yet owing to his great authority as a geometer, the mistakes which he committed as an experimentalist long became sanctioned in a world ever feted in prejudices.[5]

The relationship between mathematics and colour is a constant in many contemporary systems, yet Goethe preferred to keep his theory distinct from mathematics, only noting that geometry may be a useful basis for communication. His preference was to focus on perception and to consider the practical needs of the user. In doing so, Goethe set the context for the practically orientated text of Michel Eugène Chevreul, discussed in Chapter 3. The American painter Albert Munsell's contribution was to define a 'color notation system' in 1905, which comprehensively related 'hue, value and chroma', and established mathematical values that could permit calculation of relative colour strengths.[6] Munsell's system, presented as an 'Atlas', was based on five basic hues: red, yellow, green, blue and purple, and is still in use today.[7] The numerous colour navigation systems can be bewildering, but most employ some form of geometrical abstraction to aid comprehension and visualization.[8]

Most commonly, colour is pictured in a circle, reminiscent of the wheel produced by Newton. Although similar simple circles are still in use today, Johannes Itten, and subsequently Faber Birren, made subtle shifts to the simple colour circle in relation to perception and intensity, thus introducing irregularity into a previously balanced device.[9] In contemporary architecture, the most commonly used navigations systems in Europe and the USA are NCS, RAL, Pantone, Munsell, and British Standards.

| HUE | VALUE | CHROMA |

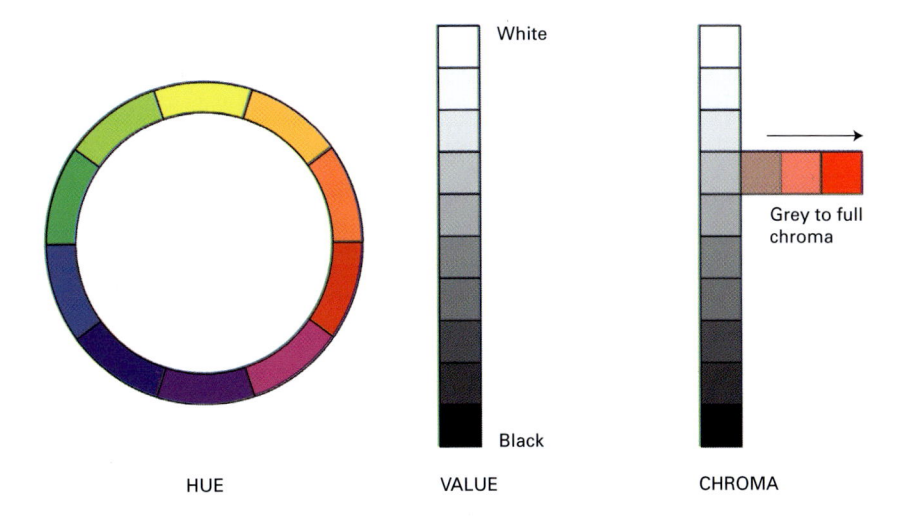

Diagram of Hue, Value and Chroma scales, after *A Color Notation*, Albert Henry Munsell (1905)

The Natural Colour System (NCS)[10] classification system is one of the most commonly used in Europe. Originated by Andres Hård, and introduced in 1979, the NCS is based on the 'Opponent Colour Theory', which was formulated by Ewald Hering in the late nineteenth century. This is represented in a three-dimensional 'colour space' or abstracted sphere. This representation is similar to Runge's *Farbenkugel* and takes a monochromatic vertical axis from white to black, with the colour hues, from red to yellow to green to blue, spanning around the widest point of the sphere. In the NCS, red and green, blue and yellow, and black and white are called opponent pairs. This system suggests a correlation between the way in which colour signals are transmitted to the brain and opponent colour theory, evidenced by our inability to perceive a colour as both reddish and greenish at the same time. Adjacent colours can be distinguished, for example a reddish-yellow or reddish-blue.

Considering that Runge also highlighted opposites in 1810, and Chevreul explored the effects of contrasts in *The Principles of Harmony and Contrast of Colours* and *The Laws of Contrast of Colour*, first published in 1839, the principles of opponent theory were largely established during the nineteenth century and have not substantially changed.

The same methodology is used in the RAL system, which first published a chart of 40 colours in 1927. It was developed in Germany, has been modified a number of times and now follows an internationally used colour measurement system, which was laid down by the Commission Internationale de L'eclairage (CIE) in 1976.[11] The hues are organized in a circle and the various designations correspond to associated angles. For example, red can be found at 0° (= 360°), yellow at 90°, green at 180° and blue at 270°. The circle is then seen at the 'equator' of an irregular sphere. The RAL system, updated in 2007, includes 1,625 colours across its full range. The most recent incarnation reduced the number of deep saturated tones and increased the number of light shades in response to market demand.

ICI/Dulux, now a subsidiary of the multinational parent Akzo-Nobel, has abandoned the NCS terminology previously used in the Dulux Dimensions range in favour of their own in-house codification. The Dulux Colour Palette has a notation giving hue, light reflectance value (LRV) and chroma.[12] The pages of their colour palette fan run in sequence from clear colour to gradual tinting with grey, and, on each page, high to low reflectance of the same hue.[13] As designers seek to address the need for colour contrast for the visually impaired, the ability to make a direct comparison of the LRV is increasingly important.

The first British Standard (BS) range of colours was published in 1930. It was limited to around 30 colours and predominately suited for external, rather than indoor, use.[14] Extended to 101 shades in 1955, it was widely used in public projects, such as schools and hospitals. To contemporary UK-based architects, the BS range gives a solid, cost-effective and easily communicated colour choice. Most paint manufacturers are able to supply the BS range, giving flexibility in the choice of supplier. It is limited, however, in certain hues and does not offer the designer the wide choice of NCS, RAL or the extended Dulux Colour Palette. As a result, it still carries an institutional association, reminding one of schools, hospitals and similar institutional buildings.

Commonality in the language of colour is essential for effective communication, and the lack of agreement on terminology and systems precludes design decision making. Other sectors have their own dialects. Pantone, founded by Laurence Herbert in 1963, predominates in the graphic arts community. The Pantone system prioritized the need to be able to match colour on the final printed surface, on coated or uncoated paper, using a fan of swatches or their Bridge, which places a Pantone colour swatch next to the effect of the same colour when processed on specific papers. Information on the CMYK (Cyan, Magenta, Yellow and Black) values is also noted. As a trademark, Pantone has become one of the best-known expressions of colour among the public, due to its use in graphic design.

Pantone colour mug,
designed by Whitbread Wilkinson
under license from PANTONE

The professional designer must learn to navigate the various systems, to understand which has the best blues, or the richest reds, although, unlike an artist, commercial realties mean that choice is frequently restricted by cost as well as by function. How then does one compare like with like?

The art of communication

BOTH NCS and RAL use the term 'colour atlas', in which abstracted slices through a three-dimensional solid are laid out. The idea that one can navigate, travel through and locate particular colours using an atlas is an enjoyable analogy.[15] Colours are ordered systematically by *hue*, *value* (lightness) and *chroma* (saturation) values. Mainly due to commercial reasons, it would seem, no single system prevails. The designer can navigate a route through the colour world, but having arrived at the preferred destination colour, trying to communicate this location to a third party is not particularly easy, unless the third party happens to speak the same language. Rather than defining colour by its technical constituents of CMYK coordinates, most architects and designers will still select colour from paper swatches or, better still, a few real samples on site where an appreciation of the final effect can be gained.

The concept of international standardization of products, materials, systems and processes developed throughout the twentieth century.[16] In 1946, following the Second World War, the British Standards Institution (BSI) was a founder member of the International Organization for Standardization (ISO)[17] and, in 1964, was a founder member of the European Committee for Standardization (CEN).[18] The ISO standards are available worldwide and provide the benchmark for communication. Nevertheless, the reality of the building industry is such that theory and practice often diverge. In 1993, the International Color Consortium (ICC) was established in the USA by a number of industrial suppliers with the aim of creating, promoting and encouraging the standardization of a neutral, cross-platform and cross-supplier colour management system for architecture and components. The outcome of this cooperation was the development of the ICC profile specification.[19]

Despite a clearly established framework for colour standardization in architecture and the building industry, Chevreul's practical concerns for the economics of errors by the dye maker, explored in Chapter 12, are still relevant. Costly errors can occur in the translation of colour choice. Contractors can be vulnerable if colours on site are not considered to match the designer's choice for various reasons and have to be corrected. The international nature of architectural commissions can be affected by the different quality of daylight, making accurate colour selection problematical. Metamerism, where colours appears to shift, can occur because of varying illumination, the angle of viewing or inherent characteristics in the observer, such as colour blindness. Is it realistic to pursue precision or standardization in something as personal, ethereal, relative and transient as colour?

In laboratory conditions it is possible to define exactly. The accurate profile or fingerprint for a colour is established using the spectrophotometric curve of the colour, where X, Y and Z

values are defined and plotted to form a geometric profile under specific lighting conditions stipulated by the International Commission on Illumination (ICI).[20] Most architects will, however, accept that variability and tolerance are a natural part of the construction industry and that complete certainty is rarely achievable. The difficulty is defining what is an acceptable range of tolerance in relation to colour deviation. As noted by Rodger Talbert in his book *Paint Technology Handbook*, 'color matching is a science which is best left to formulators and quality control experts'. As with most other elements in construction, if accuracy is required, factory conditions, and hence off-site prefabrication, will improve the ability to match colour. Pigmented materials such as ceramic tiles and fibre cement boards are also more likely to offer accuracy and quality as they are made in factory conditions. Applied coatings, however, such as coloured renders, resins and paint, will always be susceptible to the vagaries of application under site conditions. In his introduction to the *Salubra* wallpaper collection, which was the medium used for his *Claviers de couleur*, Le Corbusier noted:

> Instead of covering the walls with three coats of oils – necessarily applied amidst the hazards and hindrance arising from other work – we can now utilize this machine prepared painting and can apply it at the very last moment of finishing. The architect is always more or less at the mercy of indifferent workmanship in the matter of painting. The use of Salubra gives him peace of mind; for its proportions of oil and colour are always accurate. Its consistent quality of tone and material is guaranteed … The selection of colours no longer has to be made on the site under trying and inconvenient conditions. With 'oil paint in rolls' it is possible to select, in comfort, deliberately and with certainty, from a choice of samples which are actual pieces of the finished article.[21]

Digital colour space

WITH the migration of graphics and architectural design to digital platforms, designers have yet more problems reconciling the accuracy of colours as they appear on computer monitor screens, which are based on light, with the pigment they try to represent.[22] It seems perverse that we now need to seek translation devices to communicate effectively between what we see on screen and the pigment in its final state. Louisa Hutton, of architects Sauerbruch Hutton, has remarked, 'we have been reluctant to use the computer, partly because it is so frustrating to work with colours on the screen that look completely different once you print them out'.[23]

Research in the field of digital representation is making considerable progress in solving some of these issues. Devices such as Pantone's ColorMunki, launched in 2008, and X-Rite's Colour Eye suggest that matching, calibrating and creating colour, at least in design and printing, are becoming more predictable and consistent.[24] These digital devices can import colour from surfaces, extract colour from digital photographs and give an analysis of the colour profile. One key addition is that they can illustrate the differences in colours viewed in the internationally standard 'booth' conditions of 'North daylight', or in 'Incandescent' and 'Cool white fluorescent (CWF)' to allow a designer to visualize the relative appearance internally or externally under the most common conditions of illumination. Use of these devices in architectural design has not yet been established, but they are likely to be adopted as part of digital representation of architecture, as opposed to the built work. The use of actual samples, preferably seen on the site location and under the final lighting conditions, is still considered key and the most reliable selection method.

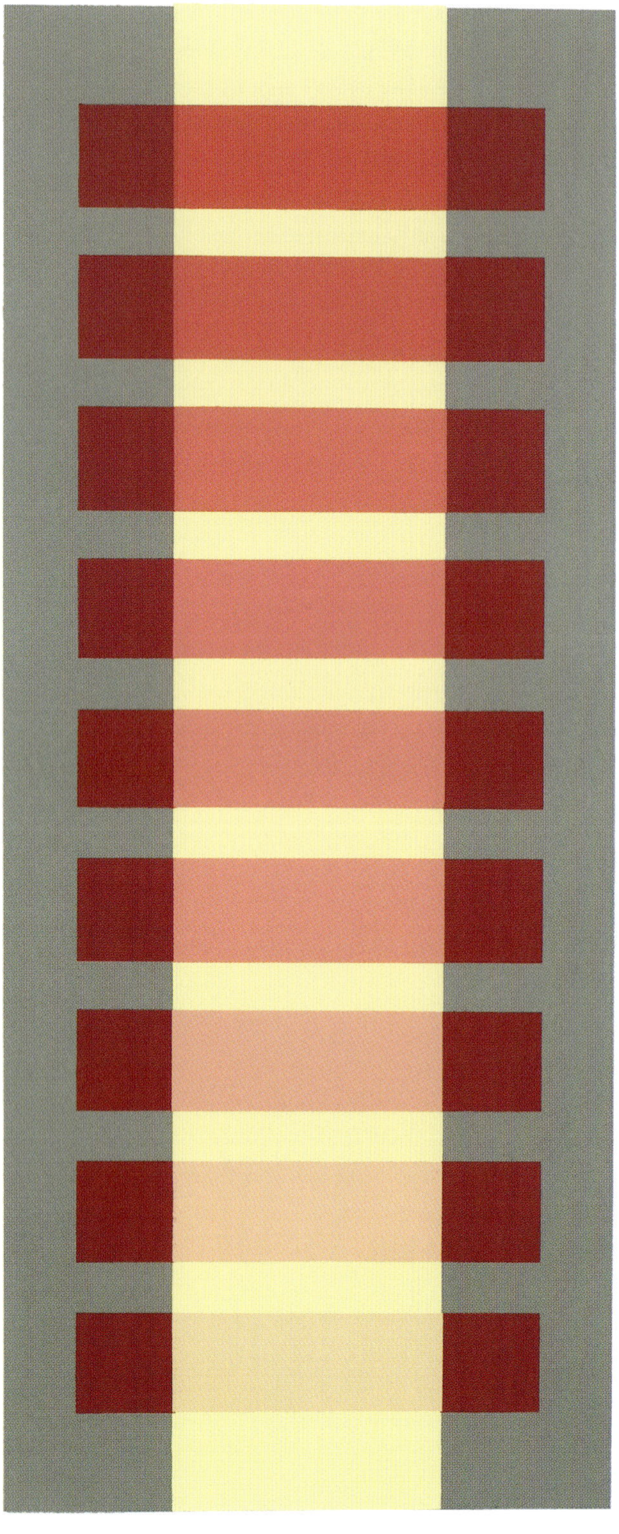

– 12 –

Playing space:
laws, rules and prescription

THE preceding chapters demonstrate a variety of approaches to the use of colour in contemporary architecture. Only a few of the architects discussed make explicit reference to colour theory, although they all have a tacit understanding of the broad principles. Architects tend to reject, or be suspicious of, systems and rigid rules that restrict design freedom yet, at the same time, acknowledge that an absence of reason is equally unsatisfactory. Paradoxically, some constraint can be liberating, imposing discipline and refinement on design decisions. Having considered the guiding principles by which contemporary architects employ colour, both intuitively and methodically, this chapter concludes by considering the role of objective theory and prescription in relation to colour design.

The *wrong* blue: subjective experience and objective recognition

To start with a personal observation – there is a particular shade of blue with a touch of green, not enough to tip it into turquoise or jade, but enough to make it appear cold, that could be considered to be 'wrong' if used architecturally. It strikes a discordant note when applied to external render and fails when used internally. It feels miserable and it sucks the colour from skin. Add some red, however, shifting it slightly towards a purply-blue, and it becomes alive, warm and wonderful. Set alongside most woods, it enhances grain, colour and natural qualities. The chromatic variation is not large, but the effect is altogether different. Is this purely subjective, or are there generally accepted architectural colours that should be avoided, restricted or their application controlled? As Johannes Itten suggests:

> above individual taste, there is a higher judgment in man, which, once appealed to, sustains what has general validity and overrules mere sentimental prejudice. This higher judgment is surely a faculty of intellect. That is why well-disciplined colour thinking

facing page
Painting from *Interaction of Color*, Josef Albers (1963)

and a knowledge of the potentialities of colours are necessary to save us from the one-sidedness and error of colouration informed by taste alone. If we can find objective rules of general validity in the realm of colour, then it is our duty to study them.[1]

Johannes Itten's teaching at the Bauhaus revolved around two recurring and opposing principles.[2] First, intuition and method and, second, subjective experience and objective recognition. The idea that there are specific rules, laws or guiding principles that govern the choice of colour pervades the entire colour industry. The abstract artist Bridget Riley finds the idea of restraint in relation to colour to be anathema. Colour for her is 'pure' rather than being the colour 'of something', and the lack of guiding principles or universal conceptual rules allows each artist to develop his or her own sensibility and unique means of expression.[3] Architects also generally consider authority and creativity as key factors in design but, unlike art, a sense of collective responsibility for the built environment will tend to constrain creativity to a greater or lesser extent. Individual colour preferences, nevertheless, do generate identifiably subjective palettes, which experience suggests are not random. In the best examples, colour is an inherent part of the conceptual design, and the chosen colours are drawn from an iterative process of critically reflective practice. Experimentation and intuition tend to dominate over rigid adherence to any sense of rules or guidance from external sources. Through his self-regulated palette, Le Corbusier argued that one could define a set of colours that would 'suffice'. As with his application of 'regulating lines' to control and order geometrical relationships, he sought to introduce discipline to colour, yet noted that 'to fix rules would be perilous'.[4]

In the 1980s, the Dutch architect Aldo van Eyck bemoaned the consistent use of 'Benthemier yellow' as demanded by the planning authority in The Hague, who justified its use on the basis of historical authenticity. Van Eyck rebelled against this academic correctness in favour of a much paler yellow, which accorded with his childhood memories of the area, but also because he simply disliked the 'correct' colour.[5] Michael Lancaster, considering the use of colour externally in urban settings, urges 'we must be careful to avoid the trap that colourfulness is a desirable objective in itself'.[6] He continues:

> guidelines are necessary to assist in understanding the ways in which colour works in the environment so that it can be used more positively and to greater effect … Although intuitive use should be accepted in a limited context, in the wider field it can easily lead to visual chaos.[7]

There is a risk that we can be so overstimulated by colour that we become visually fatigued or desensitized. There is, no doubt, validity to this position, but too much control brings with it the danger of stifling diversity in favour of a 'safety-first' attitude and monochrome monotony. In the Milos Forman film *Amadeus*, the Emperor accuses Mozart of overstimulation in this exchange:

> *Emperor Joseph II*: Too many notes.
> *Mozart*: I don't understand. There are just as many notes as I required, neither more nor less.

Colour circle, after Johannes Itten (1963). The three primary colours sit centrally in a triangle; mixing adjacent colours generates the three secondary hues, which are then surrounded by Itten's 12 main hues

Emperor Joseph II: My dear fellow, there are in fact … only so many notes the ear can hear in an evening … Your work is ingenious. It's quality work. And there are simply too many notes, that's all. Just cut a few and it will be perfect.
Mozart: Which few did you have in mind, Majesty?[8]

Laws, rules and prescription: didactic approaches

PHILIPP OTTO RUNGE, Johann Wolfgang von Goethe, Michel Eugène Chevreul and, later, Johannes Itten and Ludwig Wittgenstein suggested that there are objective principles relating to colour that should guide, although perhaps not altogether override, the subjective instinct when gauging the use of colour. Runge's *Farbenkugel* includes an illustration of simple combinations of colours labelled as 'harmonische wirkung' (harmonius effect), for example, orange and blue, yellow and lilac, and 'disharmonische Wirkung', for example, blue and yellow, yellow and red, red and blue.

Concepts of harmony, balance and contrast are rooted in the general understanding and language of architecture, not only in relation to colour, but in relation to composition in general.

In the *Laws of Contrast of Colour*, Chevreul begins by defining a long series of colour effects, before continuing to provide specific applications for his 'Laws'.[9] Chevreul lists his observations, for example: '45. Green, complementary to red, being added to white, the red appears redder and deeper' and '56. Green, the complementary of red, when placed to the side of black, makes it appear reddish. The red looks clearer, partaking less of orange.' These effects are timeless, are easily demonstrated and, one assumes, have been generally assimilated by present-day professionals, but it is not evident that they are regularly exploited in architectural contexts.

Chevreul's initial interest in colour was as a chemist, and focused on the effects of colour adjacency, particularly of dyes and textiles. He was well aware of the commercial problems that could be generated by optical illusions in cloth. He feared that if a small area of clear colour were to be set into a black or dark fabric, it might appear distorted, resulting in the rejection of reams of cloth as poorly made. If isolated, it could be proved that each of the colours was true, and that the problem was merely how the interaction of one colour with the other is perceived. He is credited with the term 'simultaneous contrast', although the effect had been known for centuries:

> When two or more contiguous colours are seen at the same time, they appear as dissimilar as possible, both with regard to their optical composition and their depth of tone. Therefore there may be at once simultaneous contrast of colour, properly so called, and simultaneous contrast of tone.[10]

The aim of Chevreul's book was to provide a series of practical experiments, which the reader was invited to try with coloured paper. He was determined to give practical application of his comprehensive laws across a wide range of activities, including the arts, dressmaking and even horticulture. His 'rules' for interiors include the relationship of furniture, hangings and wainscoting (skirtings), and are noted in relation to the effect of colours on a pale skin. He also warns of combinations of pigment against natural materials, such as wood: 'it is evident that we must assort violet or blue stuffs with yellow woods, as the root of ash, yew, satinwood, maple, etc. Violet or blue-greys are equally good with yellow woods, as green-greys are with the red woods.'[11] By continually remarking on the effect of colour on the body in his text (albeit a white Caucasian one), he aimed to make his treatise accessible to a wider audience. As discussed in Chapter 2, it was widely reported that the Impressionist painters, such as Delacroix, Monet and Renoir, were aware of Chevreul's observations and applied them to great effect.[12] Chevreul offered guidance to both sexes on the effect of colour in clothes in relation to hair and skin colour. One continues to find similar, rhetorical advice prevalent in today's popular press and on the Internet. For example, Chevreul's diagnostic advice to illustrate simultaneous contrast was that, 'A violet bonnet is always unsuitable to every complexion, since there are none [skin colours] to which the addition of yellow will be favourable'.[13] Fashion trends, particularly in clothes, will tend to overrule the idea of a personal colour palette, or of objective rules. As Chevreul lived to be over 100 years old, he would no doubt have been horrified to see the extraordinary popularity of mauve as the height of fashion in the mid-nineteenth century, only 20 years after his own text had been written.

Nearly a century later, Josef Albers adopted a similar, practical approach, but his tone is subtly different.[14] Albers' aim is also to impart understanding of optical effects, in particular the interaction of colours on each other, but once the knowledge and skills have been acquired, to encourage experimentation with these phenomena. Albers invites us to question the preconceptions of harmony, in order to equip ourselves to break the rules. He could be seen as a liberal among a series of dictators.

In the 1950s to 1970s, research in architectural and building science had a higher profile than at present. Guidance publications from the period defined specific colour palettes, as well as offering general advice on the practical application of colour in schools, factories and hospitals, and its potential for altering spatial perception. Just as John Outram has bemoaned the present ignorance of colour symbolism, much of this objective research work appears to have been forgotten by contemporary architects, or is ignored as dated, and languishes in quaint printed publications. Although detailed and serious research on colour continues to be undertaken worldwide, it rarely seems to reach the architectural profession. Yet these guidance documents provide pragmatic and logical principles that may be usefully recalled, and can be seen to have lasting validity. A 1959 publication, *Colour in School Buildings*, for example, suggests that the design of a colour scheme is 'firstly analytical and secondly a creative act of the imagination'.[15] The text argues that there needs to be cognizance of the situation; in terms of location, use and contextual conditions such as daylight and orientation, but also scope for variety and difference in support of the architectural intent. Although the guide defines a set of colours appropriate for use in school buildings and indicates where they should be applied, it is not overly dictatorial in tone. There is clearly a fine line between offering logical and practical advice, and becoming overly prescriptive. The booklet's aim was to reduce the arbitrary use of colour, or to dissuade designers from resorting to an unduly cautious approach – attitudes still prevalent today. It concludes, 'A good scheme will be one which exploits the potentialities of colour ... by a skillful use of bolder colouring, in order to be invigorating'.[16] Interestingly, certain colours tend to be associated with 'institutional' use, using the term pejoratively, conjuring up images of bland, non-controversial use of colour. These publications, however, were advocating an adventurous approach to colour in institutions, clearly aware of its potential to create stimulating environments and to promote emotional well-being.

Building typology is of great significance where objective principles of colour design are concerned and most advice is simply common sense. Hospitals and healthcare facilities have to consider the view from the patient's perspective, lying on a bed, which makes ceilings of key importance. In factories, the emphasis of research shifts to colour design as an aid to vision, with health and safety as a priority. In every case, an adjustment of *hue*, *value* and *chroma* will each play a different part in the composition.[17] Psychological effects are predominately triggered by the *hue* – from warm (reds) to cool (blue greys) – and by the degree of colourfulness of a space. The use of cool colours where machines or computers generate heat is likely to induce a more comfortable, calmer state of mind. Reflected light plays a vital part in providing work surfaces and volumes with good illumination, and so the most important factor in a working environment is the light reflectance value.[18] As far as *chroma* is concerned, the aim is to avoid producing a dull or monotonous combination by introducing a note of contrast of intensity, and possibly of hue, to one side or end of a

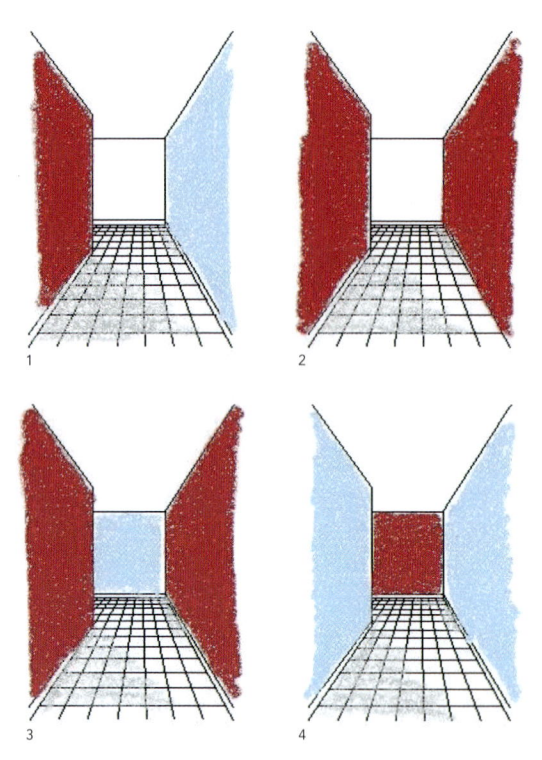

Perspective sketches showing the effect of colour altering the perception of space. Dark-toned colours in corridors tend to emphasize 'tunnel' effects as in (2); by painting one wall a pale, receding colour (1), an effect of greater width is achieved. In (3), the receding blue tint on the end wall, combined with the pinch effect from the red side walls, tends to increase the length of the corridor. Reversing the colours (4) shortens the apparent length.
Sourced from *Color in Buildings*

space, or beneath a window. Rules for domestic contexts are abundantly available, from Chevreul to Faber Birren to innumerable contemporary websites. The most consistent general principles are that stimulating colour is appropriate for circulation areas and kitchens, but not so desirable in living rooms, and may cause restlessness in bedrooms.

In addition to objective principles in relation to function in architecture, most colour theory literature is concerned with the combination and interaction of contiguous colours in relation to a composition, and the optical effects of colour on the perception of space. Complete harmony is held to be as unsatisfying as overly vibrant discord; just as in the composition of an elevation of a building, a dynamic balance is frequently more visually satisfying than a completely regular arrangement. The choice of colour and its application should be considered phenomenologically – for example to increase the apparent size of otherwise oppressively narrow or low spaces, shorten long corridors, anchor or apparently extend space, brighten dark spaces or to directly alter the mood and character of a space. Although subjective experience is impossible to regulate, all of these spatial effects can be consensually recognized and used to enhance the experience.

Colour choice and composition

THE harmony of opposing colours was central to the work of Ewald Hering in his *Opponent Colour Theory* (1870) and is embedded in many contemporary colour systems, such as the Natural Colour System (NCS, as noted in Chapter 11). Contemporary colour software offers digital harmonies of complementary, analogous and complex split harmonies based on mathematical relationships of colour values.[19] These can be considered as evidence of objective rules of general validity and may give a useful starting point for colour composition.[20] A simple example would be a large area of one hue, suitably muted by white to a tint, complemented and accentuated by a small area of a contrasting hue in full intensity. A normal expectation of the brain is that yellow is lighter than blue, and so harmony will be achieved by combining pale yellow and darker blue, whereas a dark yellow seen adjacent to a light blue may appear discordant. One should be aware, however, that colour is dynamic, and that unexpected discords can be produced even within a colour scheme that is, in theory, harmonious. The effect of daylight on colour, for example, can cause discords, particularly in interiors where light is highly directional. In a room with one window wall, the wall opposite the window tends to be brightest, and the hue shifts to a paler shade on the lower sections in direct light and darkest near the ceiling. At the same time, the window wall will appear the darkest, and the ceiling medium in tone, but lighter at the window side and darker towards the back wall. The same colour applied throughout the room will therefore appear entirely different in both colour tint and shade across the various surfaces. A pale yellow ceiling may appear darker than a blue wall bathed in intense light, and may produce unexpected and unwelcome effects.

In choosing colour, we are invited to be precise in our decisions. As architects and designers we may present our clients with a swatch, a single small piece of solid colour. In exceptional circumstances, where time and budget allow, we may prepare trial applications on site. If one simply considers the metamerism of colour under different light conditions, the idea that one can choose colour with any accuracy seems absurd. Perhaps a better mindset is to think of colour in broad bands, a sliding scale – a fuzzy perception.

Unsurprisingly, members of the general public tend to seek rules to direct the choice of colours they use in their homes and businesses – as well as their own personal appearance. Limitations and constraints are sought in response to the bewildering array of choice now confronting us in the paint store. Paint manufacturers claim authority over the public and market their advice service to emphasize the knowledge that the public do not have, or to play on our lack of confidence with colour. As noted in Dulux's 2007 advertising slogan, 'We know the colours that go', the choice of colour is seen as an activity fraught with anxiety. Such advertising perpetuates the concept that there are laws from which you deviate at your peril, as in the Dulux second slogan: 'We know the colours that don't go'. Such statements imply that the general public cannot trust their own judgement in the use of colour. Research into how the public make colour choices suggests that a complex set of parameters apply, including socio-cultural influences.[21] Notably, it is women who tend to engage in this activity more readily than men, and the marketing clearly reflects this. Paint manufacturers' websites offer personality quizzes, which identify a selection of personality traits and result in supposedly complementary colours based on the responses to the set questions. In a slightly more mootable section of his text, Johannes Itten suggests that people respond to particular colour stimuli as 'subjective timbres' and that, having made this preference overt, they should seek a career that works in equilibrium with their personality: 'A man cannot do his best except in an occupation that suits him constitutionally.' [22]

The lack of confidence in choosing and combining colours is widespread and is prevalent not only among the untrained. Professionals, in this case architects, also lack confidence in their ability to select and use colour, and are equally susceptible to the same socio-cultural influences, as well as to the suggestions of paint manufacturers and colour forecasters. There are notable exceptions, some of whom have been selected for this book. In the eight practices included, there is evidence in each case of a defined, subjective palette, and in most cases particular colours are repeated from one project to the next, as the practice gradually develops a sense of security with a particular set of hues to create a signature identification. Gigon/Guyer often adopt a strong yellow-green, Caruso St John a soft milky green, Wiesner a tempered red, warm yellow and blue, O'Donnell + Tuomey purply-blue and terracotta red, while yellows and greens are resonant of AHMM. The very shades

Dulux advertising campaign:
(top) 'We know the colours that go'
(bottom) 'We know the colours that don't go' (2007)

that Dulux insist 'don't go' are used confidently by architects Sauerbruch Hutton, who are fond of clashing pinks and oranges, derived from visual impressions of their trips to India. Their approach reflects Josef Albers' suggestion that rules are there to be broken, once a certain level of knowledge has been attained.

Josef Albers invites us to work harder at developing an eye for colour, learn to appreciate colours that we are less naturally drawn to like, and not be narrow-minded in relation to colour:

> it seems good that we are of different tastes. As it is with people in our daily life, so it is with colour. … Therefore we try to recognize our preferences and our aversions – what colours dominate in our work; what colours, on the other hand are rejected, disliked, or of no appeal. Usually a special effort in using disliked colours ends with our falling in love with them.[23]

Between pragmatism and the sublime

NEUROLOGICAL research into the parts of the brain involved in decision making suggests that the best decisions are made when reason and emotion work together. The rational brain is expected to put emotions to one side and analyse before making a decision. Plato illustrated this division using the metaphor of a charioteer, steering two horses which are pulling in opposite directions. As Jonah Lehrer summarizes: 'The Cartesian belief in reason became a founding principle in modern philosophy. Rationality was like a scalpel able to dissect reality into its necessary parts. Emotions on the other hand, were crude and primitive.'[24] Charles Blanc echoes this rupture, considering drawing to be a masculine form of art, with colour as the feminine.[25] Such absolutes, however, are divisive, and the selection and use of colour within architectural design may instead benefit from the interaction of both parts of the brain. Colour cannot be controlled, disciplined or entirely reasoned. Its essential qualities are its fluctuations, its plasticity in appearance, its ability to challenge conventional perception, to unsettle, to heighten awareness of the familiar and enhance spatial effects. It is not something to rush, or to take lightly. In making decisions on colour choice, therefore, one needs to be analytical, objective and pragmatic in relation to context, but also open to the spontaneity, delight and sublime vibrance that colour can bring. On the periphery of rules, laws and prescription, there is a space for individuality and experimentation, a conceptual playing space.

To be dictatorial in relation to something so elusive and subjective as colour seems inappropriate. Yet surely it is helpful to have a greater facility with, and understanding of, the ability of colour to add a further experiential dimension to architectural design. Ludwig Wittgenstein, in *Remarks on Colour*, considers some of the dangers of too much prescription: 'We do not want to establish a theory of colour (neither a physiological one nor a psychological one) but rather the logic of colour concepts. And this accomplishes what people have often unjustly expected of a theory.'[26] Perhaps this 'liberal' approach has the best prospect of usefulness.

Conclusion

There is good reason to develop an eye for colour. In present architectural design, we see architects defering to artists or making uneducated use of colour. This is not, however, to equate ignorance with irrationality. We are justified in inviting a degree of emotion, a degree of spontaneity into our use of colour. We need to be reminded that colour usage cannot be understood entirely in terms of colour science, but is part of a cultural and symbolic language.[27]

In the work of Steven Holl, UN Studio and Sauerbruch Hutton, colour becomes less of a rarifier and more of a perceptual and experiential modifier. Caruso St John acknowledge that they are part of a continuum and see less need for innovation and a greater need for an awareness of social and cultural associations of colour. Both they and O'Donnell + Tuomey employ colour in a traditional manner, to temper the character of a space. At the same time, both practices make intuitive and irrational choices – simply wanting to make green buildings, for example. In all these cases, however, colour is not an afterthought; even if the choice of colour is left late in the process, the idea of colour is present during the design development. Just as with the painters, the key concern for architects is not what the colour is, but what the colour does.

facing page
Colour association studies by students
at the Haus der Farbe, Zurich

Capadrutt Ernest

| jung | alt | weich | hart | krank | gesund | fett | mager | weiblich |

— ACKNOWLEDGEMENTS —

With grateful thanks to all the architects and artists interviewed for this book, and their staff, who gave freely of their time and whose work triggered and then deepened my intrigue; to Catriona Murray for her assistance with the literature review; Georgi Radev for his help with text translation; to Margaret Campbell, who kindly exported much of her colour archive from her house to mine; to Ed Brown of the Craig and Rose paint company, Fife, for an excellent lesson on paint manufacture; to architectural photographer Reinhard Görner, for acting as translator to Erich Wiesner; to Professor Iain Boyd Whyte for his invaluable editorial advice; and to the University of Edinburgh for supporting my research with an initial sabbatical semester, which gave the vital impetus to start the research project and for a research allowance, essential to fund travel and image rights.

Finally, thanks to my family – to Ewen for his encouragement, constant and enduring support and advice, and for reading and re-reading the texts, and to Max, for his patience and enthusiasm.

NOTES

Preface

1. Josef Albers' Introduction to *Interaction of Color,* New Haven: Yale University Press, 1963 (paperback, 1971), p. 1.
2. The book will use the UK spelling of 'colour' except where directly quoting an American source.
3. William W. Braham, *Modern Color/Modern Architecture*, Farnham, Hampshire: Ashgate, 2002, p. 6.

1 Introduction: investigations in the professional palette

1. Philip Ball, *Bright Earth: The Invention of Colour*, Chicago and London: University of Chicago Press, 2001, p. 193 and Faber Birren, *History of Colour in Painting,* New York: van Nostrand Reinhold, 1965, p. 13.
2. Blue Tongue Entertainment is an Australian computer game development studio. *de Blob* was first issued in 2008. The soundtrack is, incidentally, also free form and generated by the player picking up the different colours to mix the music.
3. With the possible exception of the higher instance of red/green colour-blindness in males. It is interesting to note, however, that of the eight architectural practices selected for this book, solely on the basis of the architectural output, there are five male/female partnerships. This represents a much higher prevalence of women compared to the normal make-up of the profession. Lucila Geymonat de Destefani and Allan Whitfield also report a dominance of women in domestic colour decisions in 'Esthetic decision-making: how do people select colours for real settings?' in *Colour Research and Application*, Wiley Interscience, vol. 33, no. 1, February 2008, p. 59. Perhaps an opportunity for further research, but not the focus of this book.
4. Max McLachlan (aged 11) in conversation with the author on the meaning and importance of colour, 2009.
5. Daniela Spath, *The Psychological and Physiological Effect of 'Cool Down Pink' in Human Behavior*, conference proceedings, AIC 2011 Mid-term meeting, *Interaction of Colour and Light in the Arts and Sciences*, Zurich, p. 190. The colour is being used in experiments to calm male prisoners, although, interestingly, it does not always work. Some male prisoners feel that their gender is being challenged and become more hostile; as do some women.
6. HSBC bank advert, available online at: www.visit4info.com/advert/HSBC-Gesture-HSBC-Bank/14901 (accessed November 2011).
7. Edmund Burke, *On the Sublime and Beautiful*, vol. XXIV, Part 2, The Harvard Classics, New York: P.F. Collier & Son, 1909–14; available online at: www.bartleby.com/24/2/ (accessed November 2011), and Immanuel Kant, *Critique of Judgement,* originally published 1914, New York: Cosimo, 2007. The best-known artist of the Sublime is Caspar David Friedrich, whose scenes of the terrifying power of nature frequently placed a figure with his back to the audience, looking inward in the frame. The viewer sees the scene through his eyes, but safely distanced from the actual landscape. See also Iain Boyd Whyte, 'The Sublime: an introduction', in Roald Hoffmann and Iain Boyd Whyte (eds), *Beyond the Finite: The Sublime in Art and Science*, New York: Oxford University Press, 2011, pp. 3–20.
8. Barnett Newman, *The Sublime Is Now*, 1948, in Simon Morley, *The Sublime*, Documents of Contemporary Art, London: Whitechapel, 2010.
9. E & F McLachlan Architects – the office started formally in 1990 but the two partners had been working together since they were students in the early 1980s at Edinburgh University.

10 Shane O'Toole, curator and editor, *Tales from Two Cities*, Edinburgh: Matthew Architecture Gallery, University of Edinburgh, 1994, pp. 12–13.

11 Edinburgh World Heritage Historic Home Guide, *External Paintwork*, pp. 5 & 6. Available online at: www.ewht.org.uk/property-owners-guide (accessed 24 August 2011).

12 The influence and activities of Itten, Albers, Kandinsky, Klee and Ostwald at the Bauhaus are related in Magdalena Droste, *Bauhaus, 1919–1933*, Bauhaus-Archiv, Koln: Taschen, 2002.

13 Byron Mikellides in Tom Porter and Byron Mikellides, *Colour for Architecture Today*, London and New York, Routledge, 2009, gives an outline of this research and his own work at Oxford Brookes University on colour preferences.

14 Some colours, particularly blues, are already unavailable due to European health and safety legislation restricting the use of cobalt.

15 The methodology used in the making of the painting has subsequently been developed using digital technology to study the palettes of contemporary architects, for example of AHMM in Chapter 8. The concept of specific 'tempers', or 'subjective timbres', refers to a general inclination or preference of an individual towards a group of colours.

16 Amédée Ozenfant advised a method for planning a colour scheme, which ends with the advice to '*Correct the design according to sensibility*' in 'Colour: experiments, rules, facts', one of a series of six articles *Architectural Review* 81 (April 1937) 196, reproduced as Appendix p. 111 in Braham, William W., *Modern Color/Modern Architecture*, Farnham, Hampshire: Ashgate, 2002, pp. 109–112.

2 Colour, form and material surface

1 Derek Jarman, *Chroma*, London: Vintage, 1994, p. 42.

2 Rem Koolhaas, *OMA 30: 30 Colours for a New Century*, Netherlands: Baliricum, 1999, p. 12. Koolhaas observed that: 'There are two kinds of colour. The ones that are integral to a material, or a substance – they cannot be changed – and the ones that are artificial, that can be applied and that transform the appearance of things. The difference between colour and paint.'

3 The Scottish scientist James Clerk Maxwell developed our understanding of the spectrum, light and colour recognition in the eye, and is known for his demonstration using discs of three coloured pigments which, when spun, appear to fuse into white.

4 Philip Ball gives an excellent introduction to the science and physiology of light and colour in his book *Bright Earth: The Invention of Colour*, Chicago and London: University of Chicago Press, 2001. Ball trained as a chemist, one of a number of chemists such as Michel Eugène Chevreul and Wilhelm Ostwald who have been fascinated by colour, frequently as a side-product of their main scholarly activity. His book gives useful technical descriptions of the way in which we see colour in the eye and also a detailed history of pigment in painting.

5 Michel Eugène Chevreul, *The Laws of Contrast of Colour: and Their Application to the Arts of Painting, Decoration of Buildings, Mosaic Work, Tapestry and Carpet Weaving, Calico Printing, Dress, Paper Staining, Printing, Illumination, Landscape and Flower Gardening, etc.*, translated by John Spanton, London: Routledge, 1859 (English), French edition without colour plates, 1839.

6 Johann Wolfgang von Goethe, *Farbenlehre (Doctrine of Colours)*, 1810, English translation 1840, pp. 29–37. Goethe relates the experiences of mountain climbers observing different coloured shadows as they climb – most frequently

a pale violet. He also suggests experiments in candlelight that can reproduce the effect and thus disagrees with others who believed it to be an effect of blue light in the air. He notes that the shadows will be contrasting in colour.

7   Claude Monet quoted by Philip Ball, op cit., 2001, p. 207 (J. Claretie, *La Vie a Paris*, 1881, p. 266).
8   Robert Kudielka and Richard Shone, *Bridget Riley: Dialogues on Art*, London: Thames & Hudson, 2003 (Riley talking to E.H. Gombrich).
9   Banachek, *Psychological Subtleties 2*, Houston: Magic Inspirations, 2007, p. 173.
10 From a lecture by Innes Cuthil, biologist, at *Colour in Art, Design and Nature* conference, Edinburgh, October 2008.
11 Such as 'homochromie': *pro-cryptic installations in a supermarket series*, acrylic on protective suit, available online at: http://laurentlagamba.free.fr/ (accessed November 2011).
12 Sauerbruch Hutton essay 'On colour and space' (lecture at the Gottfried Semper (1803–1879) Symposium, Athens, 2003), in Sauerbruch Hutton, *Archive*, Baden: Lars Müller Publishing, 2006, pp. 181–89, and in Tom Porter and Byron Mikellides (eds), *Colour for Architecture Today*, London and New York: Routledge, 2009, Chapter 17.
13 William W. Braham, *Modern Color/Modern Architecture*, Farnham, Hampshire: Ashgate, 2002, p. 17.
14 Arthur Rüegg's introductory text to *Polychromie Architecturale: Les Claviers de Couleurs de Le Corbusier de 1931 et de 1959*, Basel: Birkhäuser, 1998, 2006.
15 David Leatherbarrow and Mohsen Mostafavi, *Surface Architecture*, Cambridge, MA: MIT Press, 2002, pp. 91–93.
16 Ibid, p. 70.
17 Achim Borchardt-Hume, Curator Tate Modern. Lecture at Gerhard Richter symposium, National Gallery of Scotland, 21 November 2008. He noted a problem with an exit door at the Tate Modern interrupting the relationships of Richter's six 'Cage' paintings, so they had to move the door.
18 Ibid. As Richter trained as a mural artist in the Deutsche Democratic Republic, the idea of a wall as a surface capable of accommodating figurative and/or symbolic imagery would have been deeply rooted.
19 Refer to Chapter 7.
20 Robert Venturi, *Complexity and Contradiction in Architecture*, London: The Architectural Press, 1977, pp. 48, 84.
21 Rem Koolhaas, Bruce Mau, Hans Werlemann, Office for Metropolitan Architecture, *Small, Medium, Large, Extra-Large (S,M,L,XL)*, New York: Monacelli Press, 1995, p. 501. This is discussed further in McLachlan 'Dancing windows', *ARQ*, vol. 10, nos 3–4, March 2007, pp. 190–200.
22 Edi Rama, Mayor of Tirana, Albania, speaking at the Tate Modern, London 2009, as quoted in Ricky Burdett and Adam Kaasa, 'Color and the city' in *New Geographies 3: Urbanisms of Color*, Cambridge, MA: Havard University Press, 2010, p. 59.
23 John Gage, *Colour and Culture*, London: University of California Press/Thames & Hudson, 1999; John Gage, *Colour and Meaning: Art, Science and Symbolism*, London: University of California Press/Thames & Hudson, 2000; Philip Ball, op cit., 2001.
24 Prussian blue (iron ferrocyanide), for instance, which transformed the use of blue in painting, was invented in 1704. In antiquity, purple was the most revered colour, a symbol of status and importance. The dye, Tyrian purple, extracted from Mediterranean molluscs, was exorbitantly expensive due to its rarity –12,000 shellfish were sacrificed to produce 1.5 grams of the precious dye.
25 Philip Ball, op cit., 2001, p. 191.
26 The quote 'A good picture, like a good fiddle, should be brown' is attributed to Constable in Ball, op cit., 2001, p. 136. Patsy Campbell, art historian and restorer, notes that the actual discoloration being mimicked was a result of the unstable varnish. The pigments themselves were also unstable – Titian's greens, appearing as dark and sombre,

would have been bright and vibrant before the oxidation of the original and resultant copper-based verdigris (*Colour in Art and Nature* conference, Edinburgh, October 2008). William Turner, one of the first to use non-realistic touches of red on figures, may have been aware that they would not be lightfast. He would have used carmine – or cochineal – that contains the blood of insects and 'is one of the reddest dyes the natural world has produced'.

27 Philip Ball, op cit., 2001, p. 5.

28 Roger Talbert, *Paint Technology Handbook*, London: CRC Press, 2008, p. 55.

29 For example, products that are 'micro-porous' tend to have a fine molecular film on the surface that allows moisture to escape. Traditional limewash and chalky emulsions offer excellent breathability but poor durability in comparison with the hard surface generated by paint with an alkyd resin.

30 From an interview between the author and Ed Brown, Craig and Rose Ltd, Dunfermline, 22 May 2009.

31 Organic substances tend to contain carbon, whereas inorganic substances tend not to contain carbon. Water is therefore inorganic. The use of the term 'organic' to suggest a natural product is therefore confusing, as many synthetically made substances are organic.

32 Le Corbusier, *Polychromie Architecturale: Farbenklaviaturen von 1931 und 1959 [Color Keyboards from 1931 and 1959/Les Claviers de Couleurs de 1931 et de 1959]*, Arthur Rüegg (ed.), op cit., 2006, pp. 103 and 133.

33 The artists' pigment burnt sienna is made by heating yellow ochre until it loses water and turns deep brown.

34 These have largely replaced lead chromates and cadmium yellow, both of which are toxic.

35 Victoria Finlay, *Colour: Travels Through the Paintbox*, London: Hodder & Stoughton, 2002.

36 Craig & Rose, Dunfermline, supplied the iron oxide paint for the bridge maintenance for over 100 years. It has been replaced by a two-part epoxy paint, which contains glass flakes to provide a more effective barrier to the salty air. It is, however, usually applied in shop conditions, so time will tell if it will be successful in reducing maintenance requirements. It has already taken 19 years to strip back the old linseed-oil paint to bare metal. The bridge was one of the first to be made of steel, which was steeped in linseed oil for months prior to construction. The work is due to be completed in 2012.

37 Tom Porter in 'Colour in architecture' in *Architectural Design*, 1996, nos 3–4, studies colours preferences and notes an order of preference – blue, red, green, purple, orange and yellow. Also noted in Gage (1999).

38 Manufactured by companies such as Ciba-Geigy/BSF. Codes are product references that give no colour information, such as Bayferrox© 105M (a red, synthetic iron oxide alpha), or Bayferrox© 960G (a yellow synthetic iron oxide).

39 As related in Simon Garfield, *Mauve*, London: Faber & Faber, 2000, p. 105.

40 European legislation in 1994 led to the withdrawal of products dependent on cobalt (a naturally occurring ferromagnetic metal), which has been linked to various cancers. The NCS revised their colours accordingly and in *NCS Edition 2*, 261 new colours were added in 1995, 46 colours were withdrawn and about 1,000 *NCS Edition 1* colours changed slightly for reasons of pigment and accuracy. Approximately 400 *NCS Edition 1* colours were given new notations.

41 The wonderfully named Snow Lotus Biotech Co. Ltd of China is one such company, which is producing plants for dye and pigment manufacture. In Europe, paint companies such as Auro and Keim specialize in products with naturally sourced pigments.

42 For instance, white topcoat, satin matt, 'Classic, No. 936' - full declaration: mineral pigments; orange oil; mineral fillers; linseed oil stand oil; dammar; wood oil stand oil; driers (lead-free); castor oil stand oil; colophony-glycerine ester; silicic acids; swelling clay; lecithine; alcohol; water.

43 The White House in Washington (1800) was painted with lime, taking hundreds of tons to complete. Lime, if applied hot, can be hazardous, but will allow the surface to breathe.

44 After the military left the castle in the 1964, restoration of the building began but the main project did not commence until 1990. Up until this point, the Great Hall, built by James V of Scotland in 1505, had a rubble-stone facade with ashlar window surrounds.

45 The sudden change in appearance when the tarpaulin was removed caused huge controversy in Scotland, despite a pre-emptive publicity and information campaign. Stirling Castle is extremely visible, set high on a rock above the flood plain of the River Forth. Peter Buchanan, the project architect, recalls that European visitors have 'not batted an eyelid' at the pale ochre colour, whereas the local population found it difficult to accept the change. In Scotland, perhaps as a means of avoiding the regular maintenance, original renders have been stripped and external colour has become unusual.

46 Most coatings will require specification of the substrate, pre-treatment, application method, number of coats and possibly conditions for curing.

47 Readily available pigmented sheet materials such as Eternit boards can be cut and fixed as cladding, giving a solid intense colour. In factory finishes, such as powder coatings, common to metal sheeting, curing involves melting and resolidifying with resultant chemical cross-linking to form a hard, colourfast and durable surface. Rodger Talbert notes in the *Paint Technology Handbook* that testing is normal in factory applied and manufactured products, but much less common on site.

48 The product known as 3M Radiant Colour/Light Film produces interference patterns, like petrol or peacock feathers. See Niall McLaughlin's essay 'The illusive facade' in Tom Porter and Byron Mikellides (eds), op cit., 2009.

49 Schenk has held the position of artist-in-residence at the Schools of Bioscience and Physics at the University of Birmingham.

50 William W. Braham, op.cit., p. 5

51 See Chapter 11 for an overview of colour classification systems.

3   The unattainable myth of novelty: Caruso St John

1 The phrase is from a lecture by the Belgian Art Nouveau architect Henri van der Velde, 1929, who was berated by the early Modernists for his ornamented facades. Van der Velde's response was to uphold the skill of the craftsman over that of an anonymous worker, and he was supported in this by the Wiener Werkstätte architects and artists.

2 From the author's interview with Adam Caruso, London, September 2009.

3 The original building Bethnal Green Museum (now the Museum of Childhood) dates from 1872. Caruso St John undertook a refurbishment project and added a new entrance building.

4 Dan Cruickshank suggests that Soane's colour schemes, evident in the restored rooms of Pitshanger Manor (1808), were influenced by his mentor, James Peacock, who summarized his colour theories in his book *Nutshells* (1785) (written under a pseudonym, Jose Mac Packe). Cruickshank, D., 'Soane and the meaning of colour', *Architectural Review*, January 1989, vol. 185, pp. 46–52.

5 Louis Sullivan and Dankmar Adler's Guaranty Building, Buffalo (1894–95), is cited by the architects as influential in the design of Nottingham Contemporary.

6 Hal Inberg, 'Sampling and remixing an architecture of resistance' in *Canadian Architect*, vol. 44, no. 9, September 1999, pp. 26–33.

7 The term is taken from an exhibition of their work at the ETH in Zurich, 2009. The title of the chapter is taken from a text by Adam Caruso and Peter St John to accompany the 2009 exhibition of their work *Almost Everything* at

the ETH, Zurich, Switzerland. They note: '… contemporary architecture has, of late, become obsessed with formal invention and the unattainable myth of novelty. These matters have not been the focus of our work. We are first and foremost interested in the rich culture of architecture and in our work being an extension and a critique of that culture …'. Available online at: www.carusostjohn.com/practice/exhibitions/caruso-st-john-architects-almost-everything/ (accessed 26 December 2011).

8   Adam Caruso, 'The tyranny of the new' in *Blueprint*, issue 150, May 1998, pp. 24–25.

9   Refer to Victoria Finlay, *Colour: Travels Through the Paintbox*, London: Hodder & Stoughton, 2002, or Philip Ball, *Bright Earth: The Invention of Colour*, Chicago and London: University of Chicago Press, 2001, for the historical development of pigments. The technology of pigments has always had a close relationship with the history of art and architecture. Much brighter, chemically derived pigments, such as clear orange, became available early in the twentieth century and were readily adopted by artists such as Matisse.

10 Michel Eugène Chevreul, *The Laws of Contrast of Colour: and Their Application to the Arts of Painting, Decoration of Buildings, Mosaic Work, Tapestry and Carpet Weaving, Calico Printing, Dress, Paper Staining, Printing, Illumination, Landscape and Flower Gardening, etc.*, translated by John Spanton, London: Routledge, 1859 (English) French edition without colour plates 1839.

11 Jones' preface to the Folio edition of *The Grammar of Ornament* states as his intention: 'in thus bringing into immediate juxtaposition the many forms of beauty which every style of ornament presents, I might aid in arresting that unfortunate tendency of our time to be content with copying, whilst the fashion lasts, the forms peculiar to any bygone age, without attempting to ascertain, generally completely ignoring, the peculiar circumstance which rendered the ornament beautiful, because it was appropriate, and which as expressive of other wants when thus transplanted, as entirely fails' (p. 1).

12 See Victoria Finlay, *Colour: Travels Through the Paintbox*, London: Hodder & Stoughton, 2002, pp. 139–40.

13 Simon Garfield's *Mauve*, London: Faber & Faber, 2000, gives a full account of Sir William Perkin's work and the subsequent development of many other aniline dyes based on coal tar.

14 *Punch*, Saturday 20 August 1859, vol. 30, London: Punch Office, p. 81.

15 Ibid., also quoted in Simon Garfield, *Mauve*, p. 65.

16 Adolf Loos, 'Men's fashion', in *Spoken into the Void: Collected Essays 1897–1900*, translated by J. Newman and J. Smith (1982). See also Mostafavi and Leatherbarrow, *Surface Architecture* (2002), which refers to the Hoffman Palais Stoclet and its expression of thinness; even internally pilasters are flattened. Cladding is apparently freed from connection to the building.

17 Mark Wigley, *White Walls, Designer Dresses: The Fashioning of Modern Architecture*, Cambridge, MA: MIT Press, 1995, p. 115. Mark Wigley dissects the ambiguous relationship that the Modernists had with decoration, noting that colour, unlike other forms of decoration, could not be eradicated, but it could be tightly controlled.

18 Jill Stansfield and T.W. Allan Whitfield, 'Can future colour trends be predicted on the basis of past colour trends? An empirical investigation', *Color Research & Application*, vol. 30, no. 3, pp. 235–42.

19 William W. Braham, *Modern Color/Modern Architecture*, Farnham, Hampshire: Ashgate, 2002, p. 8.

20 Don Vidler, Sales Director of a textile firm quoted in Simon Garfield, op. cit., p. 72.

21 Gerhard Meerwein, Bettina Rodeck and Frank Mahnke, *Color: Communication in Architectural Space*, Basel: Birkhäuser, 2007, p. 21.

22 Rem Koolhaas, *Thirty Colours for a New Century*, Netherlands: Baliricum, 1999, p. 12.

23 Subsequent designs include galleries for Gagosian Gallery, Spike Island in Bristol and Nottingham Contemporary.

24 Refer also to R. Moore's essay, 'A pebble on water' in Caruso St John Architects, *The New Art Gallery, Walsall*, London: Batsford, 2002, pp. 58–70.

25 David Watkin, *Morality and Architecture*, Oxford: Clarendon Press, 1977, considers Pugin's comparison of architectural truth and religious truth at p. 17 and notes that, to Walter Gropius, the themes and techniques of the Modern Movement were, first, a belief in honesty, second, belief in technology and, third, belief in the spirit of youth, p. 83.

26 At their V&A Museum of Childhood, in Bethnal Green, London, some of the yellow/gold-painted soffit beams are structural steel and some faked with medium density fibreboard sheet (MDF) and, at Walsall, the depth of beams varies in relation to the scale of the room, rather than as a product of the structural loads and spans.

27 For an explanation of the source of the term 'Raumplan', refer to Kent Kleinman and Leslie van Duzer Suny, in *Architecture, Language, Critique: Around Paul Engelmann,* edited by J. Bakacsy, A. V. Munch and A.-L. Sommer, Amsterdam: Editions Rodopi BV, 2000, p. 161.

28 Beatrice Colomina's essay the 'Split wall' in Beatriz Colomina and Jennifer Bloomer's *Sexuality and Space*, New York: Princeton Architectural Press, 1992, p. 73, dissects the spatial relationships in Loos' Müller House in voyeuristic terms and in relation to the way in which rooms were defined by the gender of the users. The 'Ladies room' includes a small recess from which distant views of the garden can be seen across the main space. Colour was used by Loos to wrap the wall surfaces and the built-in furniture. Strong greens, reds, yellows and blues are all employed in addition to veneers. Pattern is only used on cloth surfaces such as the Ladies' window seat.

29 Caruso St John and Paul Vermeulen, *Knitting Weaving Wrapping Pressing*, Basel: Birkhäuser, 2002, p. 76.

30 Fiona McLachlan, 'Dancing windows', *ARQ 2006*, vol. 10, pp. 191–200. Semper's *Four Elements* (written in 1851) were: the hearth (interpreted as the moral element, the focus or meaning), the roof, the enclosure and the mound (which can be interpreted as site or location). Here, we are most concerned with the wall element which he likened to carpets hung around a hearth, then to the weaving of branches into wickerwork walls and, finally, to stucco and stone cladding. He differentiated between the 'wall fitter' and the 'mason', arguing that polychromy and the decoration of the surface was the scope of the wall fitter whereas the mason's skill could only be seen as independent in large-scale terrace walls.

31 Caruso St John and Paul Vermeulen, op cit., 2002, p. 76. Vermuelen suggests that Adolf Loos radicalized Semper's ideas by suggesting that the carpets had to be considered before the frame or supporting structure.

32 The project is reminiscent of the work of Austrian artist Oscar Putz, in the Kix Bar in Vienna, 1986, which uses flat planes of colour to disrupt the perspective and perceived spatial characteristics of the rooms.

33 Caruso notes 'it is meant to be a pure Semper façade. It's a picture of the structure so it's real architecture, but it's not the structure itself'. Interview with Paul Vermeulen in Caruso St John and Paul Vermeulen, op cit., 2002, p. 82.

34 In an exhibition of the work of Caruso St John in Zurich (2009), the architects couple key reference buildings with their own projects. In the case of Nottingham, the reference is Adler and Sullivan's Guaranty Building in Buffalo (1894–95).

35 The Victorians, who, as has been noted, embraced polychromy, also tended to use colour to give a sense of enclosure as well as to embellish surface with decoration, usually in support of form. Patterns are now re-emerging in architectural work, such as in Herzog & de Meuron's Library for the Technical School at Eberswalde, Germany (1999), normally on the exterior surface of skins used as enclosures to form. Pattern internally remains predominately in the hands of the interior designer, while architects will use colour to wrap or to define planes and elements.

36 Laura McLean-Ferris, 'Nottingham Contemporary's new building' in *Artreview.com*, 17 November 2009, available online at: www.artreview.com/forum/topic/show?id=1474022%3ATopic%3A929663 (accessed November 2011).

37 Frank Werner in *The Power of the City: Co-op Himmelblau* – he is drawing on the work of Rudolf Wittkower, Margot Wittkower and George Roseborough Collins, *Gothic Versus Classic: Architectural Projects in Seventeenth Century Italy*, London: Thames & Hudson, 1974.
38 Adam Caruso, *Frameworks* by Caruso St John, no. 8, *a+t ediciones*, July 1995, pp. 38–51.

4 An intuitive palette: O'Donnell + Tuomey

1 Tod Williams and Billie Tsien, Foreword to *O'Donnell + Tuomey: Selected Works*, New York: Princeton Architectural Press, 2007, p. 9.
2 Seamus Heaney's essay, 'Frontiers of writing', 1993, in *The Redress of Poetry*, London: Faber & Faber, 1996, p. 203, as quoted in John Tuomey, *Architecture, Craft and Culture*, Kinsale, Ireland: Gandon Editions, 2004, p. 59.
3 All quotations, unless otherwise indicated, are taken from the author's interview with Sheila O'Donnell in Dublin, 25 April 2009.
4 Tod Williams and Billie Tsien, op. cit., 2007, p. 9.
5 Tom Porter and Byron Mikellides (eds), *Colour for Architecture Today*, London: Routledge, 2009, pp. 73–77.
6 John Tuomey, op. cit., 2004, p. 57.
7 ABC Murders quoted in W. Morgan, *The Elements of Structure*, London: Pitmann, 1971, p. 18, and also in John Tuomey, op. cit., 2004, p. 57.
8 Kenneth Frampton, 'Critical regionalism: modern architecture and cultural identity' in Kenneth Frampton, *Modern Architecture: A Cultural History*, London: Thames & Hudson, 1985.
9 Tod Williams and Billie Tsien, op. cit., 2007, p. 1.
10 From the author's interview with Sheila O'Donnell in Dublin, 25 April 2009.
11 John Tuomey, op. cit., 2004, p. 11.
12 This is echoed by Sauerbruch Hutton: 'Colour is for us like a brick. You would not raise an eyebrow if someone like Louis Kahn designed another brick building. It is just because colour is being so underused at the moment that colour architecture stands out'. Matthias Sauerbruch interviewed by Aaron Betsky, reproduced in the essay 'Pleasurable and essential: colour and content in the work of Sauerbruch Hutton' in *El Croquis*, Issue 114, p. 13. See also Chapter 9.
13 Strangely Familiar was also the name given to a group of scholars and an edited book by Iain Borden, *Strangely Familiar*, London: Routledge, 1996, which considers the vitality and complexity of the everyday.
14 From the author's interview with Sheila O'Donnell in Dublin, 25 April 2009.
15 O'Donnell + Tuomey, *The Irish Pavilion*, Kinsale, County Cork: Gandon Editions, 1992, p. 24.
16 The building was under construction at the time of the interview.
17 The RAL colour system was defined in Germany in 1927, taking its acronym from the *Reichsausschuß für Lieferbedingungen und Gütesicherung* (State Commission for Delivery Terms and Quality Assurance). It was based on a collection of 40 colours under the name of *RAL 840* and has since become established as one of a number of key international classification systems for colour.

5 Who's afraid of red, yellow and blue? Erich Wiesner and Otto Steidle

1. Will Alsop oscillates between painting and architecture. In 2009, he declared that he was finished with architecture and would work solely as an artist, but this was short-lived, as he took up a consultant position shortly afterwards. Article available online at: www.guardian.co.uk/artanddesign/2009/aug/06/will-alsop-quits-architecture-painting (accessed November 2011). As quoted in Tom Porter, *Will Alsop: The Noise*, London: Routledge, 2011, pp. 157–58.
2. Mark Wigley, *White Walls, Designer Dresses: The Fashioning of Modern Architecture*, Cambridge, MA: MIT Press, 1995, p. 233.
3. Arthur Rüegg, exhibition catalogue, *Daidalos*, March 1994, pp. 66–77. Sauerbruch Hutton also makes reference to this in an essay (given as a lecture at the *Gottfried Semper (1803–1879) Symposium*, Athens, 2003).
4. Le Corbusier's *Polychromie architecturale* has been the subject of a number of books. The text was only published posthumously as part of the resurgence of interest in his *Salubra* wallpaper series (1931 and 1959). The painter Fernand Léger was influential during this period as was Alberto Sartoris, who distinguished between 'neo-plastic' and 'dynamic' use of colour. The debate also suggested a difference between the 'concrete', solid and softer tones, as used by Le Corbusier and prevalent in France, and primaries more often used in Germany and the Netherlands. Jan de Heer's book *The Architectonic Colour: Polychromy in the Purist Architecture of Le Corbusier*, Rotterdam: 010 Publishers, 2009, traces Le Corbusier's colour use back to a formative period from 1918 to the early 1920s. *Le Poème de L'Angle Droit* is a series of paintings and poems from 1947–53.
5. Bruno Taut, writing to his brother Max Taut, 2 March 1902, in Iain Boyd Whyte, *Bruno Taut and the Architecture of Activism*, Cambridge: Cambridge University Press, 1982, p. 20.
6. Ibid., p. 31. The association of colour with the social agenda may have been due to post-rationalization on the part of the architect.
7. Otto Schily in Winfried Brenne, *Bruno Taut: Master of Colourful Architecture in Berlin,* Berlin: Braun, 2008, pp. 10–11.
8. Werner Graeff reporting the remark in Karen Kirsch, *The Weissenhofsiedlung Experimental Housing for the Deutscher Werkbund, Stuttgart 1927*, New York: Rizzoli, 1989, p. 137.
9. The Villa La Roche (1923) contradicts this approach. Le Corbusier changed colours on different planes within the same space, evidently more interested in the relationship between colour and formal elements than in unity. Le Corbusier's views on colour fluctuated throughout his life and he clearly never entirely settled on a prescriptive method, despite his highly restricted palette of colours.
10. Deborah Gans, *Le Corbusier Guide*, New York: Princeton Architectural Press, 2006, p. 130.
11. Alfred Roth subsequently wrote on the relationship between painting and architecture, *Architektur und Malerei*, drawing heavily on Theo van Doesburg's work and contradicting Le Corbusier's position. His text is thought to have been a response to a lecture by Fernand Léger in 1933, with which he did not agree.
12. Theo van Doesburg's essay 'The dwelling: the famous Werkbund exhibition' originally published in *Het Bouwbedrift*, vol. 4, no. 24, November 1927, pp. 556–59, republished in Charlotte and Arthur Loeb (eds), *Theo van Doesburg on European Architecture: Complete Essays from Het Bouwbedrift 1924–1931*, Basel: Birkhäuser, 1990, p. 170.
13. Bruno Taut, *Ein Wohnhaus*, Stuttgart: Franckh'sche Verlagsbuchhandlung W. Keller & Co., 1927, p. 90.
14. Andrew Mead, 'Colour vision' in *Architects' Journal*, 11 October 2001, vol. 214, no. 13, pp. 34–43.
15. Dietmar Elger (ed.), *Donald Judd Colorist*, Stuttgart: Edition Cantz, 2000, p. 17.
16. A number of the architects interviewed for this book, even those who do not collaborate with an artist, cited their interest in painting, both as a way of studying colour and as a direct source for colour combinations that can be adopted.

17 Bridget Riley, 'Colour for the painter', in Trevor Lamb and Janine Bourriau (eds), *Colour: Art and Science*, Cambridge: Cambridge University Press, 1995, p. 31.
18 Robert Kudielka, 'Nothing by appearance', 1978, in *Robert Kudielka on Bridget Riley*, London: Ridinghouse, 2005, p. 41.
19 Ibid., p. 76.
20 Artists such as Ellsworth Kelly, Andy Warhol, Sigmar Polke, Frank Stella, Yves Klein and Dan Flavin.
21 David Batchelor, *Chromaphobia*, London: Reaktion Books, 2000.
22 John Ruskin as quoted in John Hyman, *The Objective Eye: Colour Form and Reality in the Theory of Art*, Chicago and London: University of Chicago Press, 2006, p. 45.
23 'Art as technique' in *Russian Formalist Criticism: Four Essays*, translated by Lee T. Lemon and Marion J. Reis, Omaha: University of Nebraska Press, 1965, p. 12.
24 Herschel Browning Chipp and Peter Howard Selz, *Theories of Modern Art: A Source Book by Artists and Critics*, Berkeley, CA: University of California Press, 1968, p. 548.
25 Fiona McLachlan and Richard Coyne, 'The accidental move: accident and authority in design discourse', in *Design Studies*, no. 22, 2001, p. 91.
26 Steidle Architekten is a Munich practice established in 1969 by Otto Steidle, who died in 2004. It was previously known as Steidle + Partner.
27 From the author's interview with Johannes Ernst, partner in Otto Steidle Architekten, Munich, June 2010.
28 William W. Braham, *Modern Color/Modern Architecture*, Farnham, Hampshire: Ashgate, 2002, p. 46.
29 Erich Wiesner, comment translated by the Berlin photographer Reinhard Görner, in an email to the author, 24 August 2011.
30 From the author's interview with Johannes Ernst, partner in Otto Steidle Architekten, Munich, June 2010.
31 Florian Kossak (ed.), *Otto Steidle Structures for Living*, Zurich: Artemis, 1994, p. 122.
32 The project referred to was Olympic Village in Turin (2006) in the author's interview with Erich Wiesner, Berlin, May 2009.
33 Florian Kossak (ed.), op. cit., 1994, pp. 141–42. (Wiesner has noted to the author latterly that he thinks there were only 14 colours in the final palette.)
34 From the author's interview with Erich Wiesner, Berlin, May 2009.
35 Yves Klein was disturbed by the fact that the pure pigment colour lost some of its vibrancy when mixed with the agent and sought out a way of maintaining the colour. He collaborated with Edouard Adam, a paint retailer, in 1955 to create his own blue and his first exhibition *Proclamation of the Blue Epoch* subsequently included 11 monochrome paintings in the new blue (1957) Rhodopas M60A, manufactured by the Rhone-Poulenc chemicals company and patented as International Klein Blue in 1960 – 'attracted by an idea of blueness'. See Philip Ball, *Bright Earth: The Invention of Colour*, Chicago and London: University of Chicago Press, 2001, pp. 2 and 280.
36 From the author's interview with Johannes Ernst, partner in Otto Steidle Architekten, Munich, June 2010.
37 Oliver Hamm, 'From Genter Straße to Freischützstraße: residential buildings by Steidle + Partner' in *Steidle + Partner Wohnquartier Freischutstraße, München*, Stuttgart: Axel Menges, 2003, p. 17.
38 Rudi Fuchs, 'Master of color' in *Donald Judd, The Moscow Installation* (cat) Galerie Gmurzynska, Cologne, 1994, p. 11.
39 The Turin project is notable for the freedom experienced by the architectural team. Johannes Ernst of Steidle Architekten remembers that the city was apparently so preoccupied with the preparations for the Olympics that

there was almost no control on the part of the client. The colours for the Olympic village were derived, in part, from historical colours of the centre of Turin, but with a shift in intensity to make them more lively and contemporary. In all, there are 11 colours distributed over the 80 facades of the village.

40 According to Johannes Ernst, from the author's interview with him, Munich, June 2010.

41 There is no doubt that, within the restricted budgets invariably applied in social housing, colour is commonly used to create a sense of place and a social, as well as a physical, identity. Taut's 1919 manifesto, *Die Erde eine Gute Wohnung*, also touched on this fact: 'Colour is zest for life, and because it can be given with little means we must push for it in this time of need'.

42 Oscar Putz's Kix Bar Vienna is illustrated in Chapter 2.

43 Le Corbusier and Amédée Ozenfant's essay on *Purism*, in the fourth issue of *L'Esprit Nouveau*, 1921.

6 Place, space, colour and light: Steven Holl

1 Otto von Simpson, *The Gothic Cathedral*, New York: Pantheon, 1956, pp. 51–52.

2 Le Corbusier's design for the Church at Firminy was completed posthumously, after a prolonged development period.

3 Ulrich Bachmann, *Farbe und Licht/Colour and Light*, Zurich: Verlag Niggi AG (book and DVD-rom), 2011.

4 Faber Birren, in *Light, Color and Environment: A Discussion of the Biological and Psychological Effects of Color, With Historical Data and Detailed Recommendations for the Use of Color in the Environment*, New York and London: Van Nostrand Reinhold, 1982, p. 9, explains 'Metamerism' in this way: 'where one color looks different in certain light conditions. Color needs to be chosen in the type of light source in which it will be used.' The effect usually applies to dyes, less so to pigments, and so is more important in textile design.

5 Josef Albers, *Interaction of Color*, Yale University Press: New Haven, 1963, pp. 27 and 71–72.

6 Alberto Gomez-Perez in the introduction to Steven Holl's book *Intertwining*, Princeton: Princeton Architectural Press, 1994, p. 12.

7 Steven Holl's use of patterned glass may also make reference to Eero Saarinen's leaded glass designs in his father's building on the campus.

8 A palette also used in his Y House, Catskill Mountains (1997–99).

9 Steven Holl, taken from his lecture for the Jencks Award 2010 at the RIBA, London. 'Spatial energy' he noted as one of his five 'axioms' on architecture.

10 Doreen Balabanoff, *International Color Association (AIC) Conference* proceedings, June 2011, p. 49. The best-known example of the movement of sunlight throughout the day and year is provided by the Pantheon in Rome.

11 This is a direct observation by the author from her experience of teaching first-year architecture students. Although thankfully rare, some students coming into higher education do not seem to know which direction the sun moves around the sky.

12 Kenneth Frampton, 'Towards a critical regionalism: six points for an architecture of resistance,' in Hal Foster (ed.), *The Anti-Aesthetic*, Seattle: Bay Press, 1983, pp. 20–21.

13 Dominique and Jean-Philippe Lenclos, *Colors of the World: The Geography of Color*, New York: W.W. Norton & Company, English edition 2004, original 1999, Paris: Groupe Moniteur.

14 Paper given by David Batchelor at *Colour in Art and Nature* conference, Edinburgh, October 2008.

15 Jürg Rehsteiner, Lino Sibillano and Stefanie Wettstein, *Farbraum Stadt: Box ZRH Das Buch Eine Untersuchung und*

ein Arbeitswerkzeug zur Farbe in der Stadt, Zurich: Haus der Farbe, CRB, Kontrast, 2010, in collaboration with NCS Colour Centre Schweiz, p. 54.
16 Barragán acknowledges the inspiration of the paintings of Chucho Reyes as well as those of de Chirico, Delacroix and Magritte. He searches for colours in paintings that match his own imagined colours.
17 Szyszkowitz and Kowalski were key members of the *Grazer Schule* (Graz School), as defined in a seminal catalogue of architects based in the South Austrian town in the early 1980s.
18 Where Szyszkowitz and Kowalski differ greatly from Holl is that they tend not to write about their own work, leaving the interpretative task to commentators and, for that reason, are perhaps less well-known or understood.
19 Kenneth Frampton, 'On the architecture of Steven Holl' in Steven Holl, *Anchoring*, New York: Princeton Architectural Press, 1989, p. 8.
20 Steven Holl in response to questions at the Charles Jencks Award Lecture, RIBA, London, December 2010.
21 Todd Gannon (ed.), *Steven Holl, Simmons Hall*, Princeton Architectural Press, New York, 2004, p. 77. 'Perfcon' is the name given to a precast concrete panel system designed for the building with engineer Guy Nordenson and manufactured in Canada. Each of the 6,000 panels was unique, either in form or in the reinforcement.
22 Grete Smedal, 'The Longyearbyen Project: approach and method', in Tom Porter and Baron Mikellides (eds), *Colour for Architecture Today*, London: Routledge, 2009, Part 3, Chapter 13. Refer also to the case study section of the NCS website, available online at: www.ncscolour.co.uk/ (accessed November 2011).
23 Faber Birren, op cit., 1982, p. 9.

7 Surface and edge: Gigon/Guyer

1 Christian Norberg-Schultz, *Genius Loci: Towards a Phenomenology of Architecture*, London: Academy Editions, 1980 p. 5.
2 Charles Rattray and Graeme Hutton, 'Concepts and material associations in the work of Gigon/Guyer' in *ARQ*, vol. 4, no. 1, 2000, p. 20.
3 Ibid., p. 21.
4 Adrian Forty, 'The material without a history' in Jean-Louis Cohen and G. Martin Jr Moeller (eds), *Liquid Stone: New Architecture in Concrete*, Basel: Birkhäuser, 2006, p. 35.
5 J. Christophe Bürkle and Monica Landert (eds), Annette Gigon and Mike Guyer, *Gigon/Guyer: Works and Projects 1989–2000*, Zurich: Berlag Niggli AG, 2000, p. 139.
6 Martin Steinmann, 'Conjectures on the architecture of Gigon/Guyer' in J. Christophe Bürkle and Monica Landert (eds), op. cit., p. 214.
7 Annette Gigon, 'Materials and colours' in Petra Hagen Hodgson and Rolf Toyka, *The Architect, the Cook and Good Taste*, London: Birkhäuser, 2007, pp. 38–49.
8 J. Christophe Bürkle and Monica Landert (eds), op. cit., p. 116.
9 David Leatherbarrow and Mohsen Mostafavi, *On Weathering: The Life of Buildings in Time*, Cambridge, MA: MIT 1993/1997, 2nd edition, p. 42.
10 Wilfrid Wang, 'The variegated minimal' in *El Croquis*, no. 102, 2000, p. 27.
11 Wang, ibid., p. 25.
12 Foreign Office Architects, *Phylogenesis*, FOA, Barcelona: Actar, 2004, pp. 7–8, quoted in J. Rendell, *Art and Architecture: A Place Between*, London: I.B. Tauris, 2006, p. 68.

13 The exhibition 'Off the wall' (2007–08) at the Indianapolis Museum of Art can be viewed at www.imamuseum.org/adrianschiess/ (accessed November 2011).
14 *El Croquis*, no. 143, 2009 (interview with Adam Hubertus Stanislaus von Moos), pp. 12–13.
15 From the author's interview with Annette Gigon, Zurich, 26 June 2009.
16 *El Croquis*, no. 143, 2009 (interview with Adam Hubertus Stanislaus von Moos), p. 13.
17 Max Wechsler's essay 'Beauty is admissible: architecture as visual event' in J. Christophe Bürkle and Monica Landert (ed.), op. cit., pp. 364–65.
18 Harald F. Müller, quoted by Wechsler in J. Christophe Bürkle and Monica Landert (eds), op. cit., p. 365.
19 Gigon in Hodgson and Toyka, op. cit., p. 44.

8 Memories, associations and the brightness of yellow: AHMM

1 From the author's interview with Simon Allford, partner in AHMM, London, January 2009.
2 A collaboration with the artist Martin Richmond. The school teaches a number of profoundly deaf students and others with special educational needs.
3 From a lecture by Paul Monaghan in the Architecture Foundation series 'REAL architecture' 13 November 2008. Available online at: www.architecturefoundation.org.uk/programme/2008/real-architecture/paul-monaghan-ahmm-presents-westminster-academy (accessed November 2011).
4 Kevin Lynch, *Image of the City*, Cambridge, MA: MIT, 1960, p. 3.
5 Kenneth Boulding, *The Image: Knowledge in Life and Society*, Michigan: University of Michigan Press, 1961, p.6.
6 Ella Chmielewska, *Implaced Communication; Wayfinding and Informational Environments*, 2001, Department of Art History and Communication Studies, McGill University, Montreal, p. 268. Available online at: http://digitool.library.mcgill.ca/R/?func=dbin-jump-full&object_id=36891&local_base=GEN01-MCG02 (accessed November 2011).
7 John Outram, in Tom Porter and Byron Mikellides (eds), *Colour for Architecture Today*, London: Routledge, 2009, pp. 103, 105.
8 The ceiling is named after the donors Sue and Steve Shaper, who funded the scripted ceiling. It is vaulted and covers an area of 22 x 16 m. John Outram painted a freehand A1+ sized watercolour with imagery drawn from his own iconography. This was photographed, scanned at a very high resolution and effectively enlarged 32 times. In Outram's office, further computer software was used by Anthony Charnley to produce long, narrow sections of the images as 234 tiled pieces. Source www.johnoutram.com/projectsmenu.html (accessed November 2011).
9 Comment by Paul Monaghan, Architecture Foundation lecture at Westminster Academy, November 2008.
10 Comments from the project architect to the author, June 2009.
11 Steve Rose, *The Guardian*, 7 January 2008, available at: www.guardian.co.uk/politics/2008/jan/07/architecture.newschools (accessed November 2011).
12 Email to the author from Morag Myerscough, August 2009. Project architect comment in notes sent to the author, June 2009.
13 Monaghan served on the Centre for the Built Environment (CABE) panel, which reviewed school design through a system of design-quality indicators in 2006.
14 Studio Myerscough have worked independently on the renovation of the Barbican Arts Centre in London. To aid recognition and understanding within this complex building, where people were known to get lost, a wayfinding

system was introduced. An orange colour (a colour used on the exterior of the main hall which was part of the original architect's scheme, and worked well with bush-hammer concrete) is used throughout as a clear marker for the visitor to identify and follow. Myerscough comments that, in other instances, when wayfinding might want to be more discreet and not shout out, other techniques can be used, such as embossing walls or using fret-cut materials rather than introducing colour.

15 Romedi Passini, *Wayfinding in Architecture*, New York: Van Rostrand Reinhold, 1984, pp. 64, 66.
16 Ibid., pp. 90, 91.
17 From a lecture by Paul Monaghan in the Architecture Foundation series 'REAL architecture', 13 November 2008. Available online at: www.architecturefoundation.org.uk/programme/2008/real-architecture/paul-monaghan-ahmm-presents-westminster-academy (accessed November 2011).
18 Jean-Philippe and Dominique Lenclos, *Colors of the World: The Geography of Color*, New York: W.W. Norton & Company, 2004. This is discussed in more detail in Chapter 6.
19 As opposed to the gloominess of some of the painter Mark Rothko's use of dark crimsons, browns, greys and blacks in his series for the Seagram building in New York (1958–59), which Allford also appreciates for the way in which the intense colour invites the viewer into the depth of the image, but which would not, in his opinion, be appropriate in architectural settings.
20 Johannes Itten is likely to have made reference to the work of Wilhelm Ostwald, who used mathematical relationships to define gradations in scales of colour. Ostwald's texts were used at the Bauhaus and he also set down a series of 'laws' in relation to colour harmony.
21 Refer to Philip Ball, *Bright Earth: The Invention of Colour*, Chicago and London: University of Chicago Press, 2001, Chapter 2, for a full explanation.
22 Faber Birren, *Principles of Color*, New York: Van Nostrand Reinhold, 1969, pp. 24–25. Birren reasons that warm colours are used more frequently by artists and designers because of their more dynamic qualities. Refer also to www.colorsystem.com (accessed November 2011) for descriptions of the most notable colour systems.
23 Rem Koolhaas, Bruce Mau, Hans Werlemann, Office for Metropolitan Architecture, *Small, Medium, Large, Extra-Large (S,M,L,XL)*, New York: Monacelli Press, 1995, p. 1158.
24 See www.ral-farben.de/ (accessed November 2011).
25 Such as ColorPicker by Forged at www.albinoblacksheep.com/text/picker (accessed November 2011) or the UltraColour system that has been developed jointly by Angela Wright and the Colour & Imaging Institute at the University of Derby, England.
26 Paul Finch in Iain Borden, *Manual: The Architecture and Office of Allford Hall Monaghan Morris*, Basel: Birkhäuser, 2003, p. 55.
27 From the author's interview with Simon Allford, London, January 2009.

9 Synergies and discords: Sauerbruch Hutton

1 Matthias Sauerbruch interviewed by Aaron Betsky, reproduced in the essay 'Pleasurable and essential: colour and content in the work of Sauerbruch Hutton', *El Croquis*, Issue 114, p. 13.
2 Kieran Long, 'I just wish that Sauerbruch Hutton's decorative virtuosity meant more than some observed colours from the surroundings arranged into an attractive tiled pattern', *Architects' Journal*, vol. 229, no. 6, 19 February. 2009, pp. 26–37.

3   Louisa Hutton acknowledged that there are key staff in the office who also have a strong sensibility and are similarly tuned. (From the author's interview with Louisa Hutton, Berlin, May 2009.)
4   Robert Harbison, 'Much still to learn about using colour' (exhibition review) in *Architects' Journal*, vol. 197, no. 12, 24 March 1993, p. 52.
5   Alvin Boyarski was Chair of the Architectural Association from 1971 until his death in 1990.
6   Sophie Lissitzky-Kuppers, *El Lissitzky: Life, Letters, Texts*, London: Thames & Hudson, 1980, p. 21.
7   Alberto Gomez-Perez and Louise Pelletier, *Architectural Representation and the Perspective Hinge*, Cambridge, MA: MIT Press, 1997, 2000.
8   Mark Wigley, *White Walls, Designer Dresses: The Fashioning of Modern Architecture*, Cambridge MA: MIT Press, 1995, p. 205.
9   Sauerbruch Hutton, *WYSIWYG*, London: Architectural Association, 1999, p. 21. Refer also to Sauerbruch Hutton, *Archive*, Baden: Lars Müller Publishers, 2006, for projects such as the Junction Building, Birmingham (1989) and Tokyo International Forum (1989).
10 Sauerbruch Hutton, 'The landscape garden as a paradigm in urban design' (first published in *Arch + 118*, September 1993), republished in Sauerbruch Hutton, *Archive*, op cit., 2006, pp. 39–41.
11 Hutton notes that the clients have since changed the 'Mexican pink' colour of the pink wall to a burnt orange, but only after consulting them. In all other respects they are happy to retain the colour schedule. (From the author's interview with Louisa Hutton, Berlin, May 2009.)
12 Josef Albers, *Interaction of Color*, New Haven and London: Yale University Press, 1963 (revised 2006).
13 Kurt Forster, in Sauerbruch Hutton, *WYSIWYG*, op. cit., pp. 7 and 72.
14 Ibid., p. 62.
15 Mattias Sauerbruch, lecture at the *Gottfried Semper (1803–79) Symposium*, Athens, 2003, reprinted in Tom Porter and Byron Mikellides (eds), *Colour for Architecture Today*, London: Routledge, 2009, p. 95.
16 Jean-Philippe and Dominique Lenclos, *Colors of the World: The Geography of Color*, English edition, 2004, New York: W.W. Norton & Company, original French edition 1999, Paris: Groupe Moniteur.
17 Hendrik Petrus Berlage was a Dutch architect who studied under Semper (1856–1934).
18 This is discussed further in David Leatherbarrow and Mohsen Mostafavi, *Surface Architecture*, Cambridge, MA: MIT, 2002, p. 102.
19 Such is the present proliferation of irregular facades that Russell Jones has remarked: 'In London today the ubiquitous shuffled facade I guess is the work of lazy architects unwilling to reap the benefits of architecture considerate of context, history, structure, materiality and function. Part of what may be construed as laziness, is also a sense that architects are simply minimizing risk, knowing that the playful facade is likely to secure planning permission, while allowing an element of expression that a "dumb" plan will not readily generate.' *ARQ*, vol. 11, no. 1, 2007, p. 5 (letters).
20 Gerhard Richter, *Text*, 1961–2007, Köln: Verlag der Buchhandlung Walther König, 2008.
21 Fiona McLachlan, 'Dancing windows', *ARQ*, vol. 10, no. 3/4, March 2007, cover of journal and pp. 190–200. Journal article by the author considers irregularity in the contemporary facade.
22 Given nicknames such as 'the baguette' and 'the pill box' and supporting the different identities. (From the author's interview with Louisa Hutton, Berlin, May 2009.)
23 Available online at: www.ncscolour.co.uk/ (accessed November 2011).
24 S. Westland, K. Laycock, V. Cheung, P. Henry and F. Mahyar, 'Colour harmony' in *Colour: Design & Creativity*, vol. 1, no. 1, 2007, pp. 1–15.

25 David R. Hay, *The Laws of Harmonious Colouring: Adapted to Interior Decorations, Manufacturers and Other Useful Purposes*, London: W.S. Orr, Edinburgh: W. & R. Chambers, 1828, p. 25.

26 The colour scheme by E & F McLachlan Architects (the author) for the redecoration of the Reid Concert Hall in Edinburgh (1993) made reference to the theories of David Hay. A general tone, key or family of colours were chosen from the NCS range (in this case R70B, a purply-blue) and then, as suggested by Hay, a high note, or contrasting colour, was used to enliven and give slight discord. Professor John Donaldson, who commissioned the original building by architect David Cousin, finished in 1859, was (with Hay) one of the founding members of the Aesthetic Club of Edinburgh. Being able to justify the colour choice as an interpretation of Hay's text was clearly not lost on the academic client group at the time of the redecoration.

27 Michel Eugène Chevreul, *The Laws of Contrast of Colour: and Their Application to the Arts of Painting, Decoration of Buildings, Mosaic Work, Tapestry and Carpet Weaving, Calico Printing, Dress, Paper Staining, Printing, Illumination, Landscape and Flower Gardening, etc.*, John Spanton (transl.), London: Routledge, 1859 (English edition), French edition without colour plates, 1839. It is not evident whether Hay would have been aware of Goethe's text *Die Farbenlehre* 1810, *Theory of Colours*, English edition, London: Murray, 1840.

28 Newton's identification of seven colours was scathingly criticized by Goethe in his *Theory of Colours,* op.cit. It has been suggested that Newton was all too readily trying to make connections with cosmology and could not see that his named colours were an arbitrary selection when the reality is a continuous range.

29 Robert Venturi, *Complexity and Contradiction in Architecture*, London: The Architectural Press, 1977, p. 41.

30 The artist Bridget Riley, for instance, has noted that Poussin also linked colour to music in relation to harmonies and scales. Bridget Riley, 'Colour for the painter' in Trevor Lamb and Janine Bourriau (eds), *Colour: Art and Science*, Cambridge: Cambridge University Press, 1995, p. 46.

31 Wassily Kandinsky (1866–1944) worked at the Bauhaus from 1922 until it was closed by the Nazis in 1933. See Kandinsky, 'Concerning the spiritual in art' (1911), (M.T.H. Sadler, transl.), reprinted Las Vegas: IAP, 2009, and in W. Braham, *Modern Color/Modern Architecture*, Farnham, Hampshire: Ashgate, 2002, p. 25.

32 This is further explored in Stephen Whittle, 'Uncertain harmonies', in *Colour: Design & Creativity*, 2007, vol. 1, no. 1, 10, pp. 1–4.

33 Ibid., p. 3.

34 The physical standard representing the metre was to be constructed so that it would equal one ten-millionth of the distance from the North Pole to the equator along the meridian running near Dunkirk in France and Barcelona in Spain. It was later found that Delambre and Mechain, the surveyors, had not properly accounted for the flattening of the surface of the earth as it nears the equator in making their calculations. See http://lamar.colostate.edu/~hillger/origin.html (accessed 16 November 2011).

35 Le Corbusier, *The Modulor: A Harmonious Measure to The Human Scale Universally Applicable to Architecture and Mechanics*, originally published in 1950, London: Faber & Faber, 1951, pp. 130–131. Colin Rowe, *The Mathematics of the Ideal Villa and other Essays*, Cambridge, MA: MIT, 1976, pp. 2–16 and pp. 159–83.

36 From a conversation with Clive Albert, an architect who undertook a measured survey of Le Corbusier's projects in India. Balkrisna Doshi, who was project architect at the time, reported to Albert that Le Corbusier was in the habit of drawing over the dimensioned grid with a pencil, making final adjustments.

37 Josef Albers, op. cit., p. 41.

38 Johann Wolfgang von Goethe, *Theory of Colours*, op. cit., p. 300.

39 Andreas Rottenschlager, 'The sound of the cyborg' in *The Red Bulletin*, March 2011. See www.neilharbisson.com/ (accessed November 2011).
40 Stephen Whittle, op. cit., pp. 1–4.
41 Amédée Ozenfant, 'Colour: experiments, rules, facts', in *Architectural Review*, vol. 81, April 1937, p. 196. Part of a series of articles on colour that he wrote for this journal.
42 Josef Albers, op. cit., p. 46.
43 Mattias Sauerbruch, as quoted in *WYSIWYG*, op. cit., p. 14.
44 Arthur Rüegg, *Polychromie Architecturale: Les Claviers de Couleurs de Le Corbusier de 1931 et de 1959*, Basel: Birkhäuser, 1998, 2006, p. 69.
45 Available online at: www.museum-brandhorst.de/ (accessed November 2011).
46 Johannes Itten and Faber Birren, *The Elements of Color: A Treatise on the Color System of Johannes Itten, Based on his Book 'The Art of Color', 1961*, London: John Wiley & Sons, 1970, p. 54. Johannes Itten (Swiss), 1888–1967, worked in Berlin and Vienna then came to the Bauhaus where he taught between 1919 and 1923. He founded his own school in Berlin in 1926 (named the Itten School after 1929), which closed in 1934. Source: Magdalena Droste, *Bauhaus, 1919–1933*, Bauhaus-Archiv, Koln: Taschen, 2002, p. 246.

10 Transformational, instrumental colour: UN Studio

1 Available online at: http://zahahadidblog.com/interviews/2007/06/08/interview-with-patrik-schumacher (accessed November 2011).
2 The critic Hugh Pearman notes that the original intention was to use shiny titanium as a contrast to the concrete, but that the black aluminum was a cost-saving measure. See online at www.hughpearman.com/articles4/hadid4.html (accessed November 2011). A rare use of colour is seen in their Glasgow Transport Museum (2011), where a limey yellow-green is used in the interior. The implications of using green in sectarian Glasgow are yet to be seen, green being firmly associated with the originally Catholic football team Celtic.
3 United Network Studio (UN Studio), formerly van Berkel and Bos.
4 Kristin Feireiss in Ben van Berkel, *Mobile Forces/Mobile Kräfte*, Kristin Feireiss (ed.), Berlin: Ernst & Sohn, 1994, p. 14.
5 Todd Gannon, *UN Studio: Erasmus Bridge, Rotterdam, The Netherlands*, New York: Princeton Architectural Press, 2004, p. 16.
6 Ben van Berkel and Caroline Bos, 'How modern is Dutch architecture?' in *Delinquent Visionaries*, Amsterdam: Ben van Berkel, 1993.
7 From the author's interview with Grainne Hassett, Dublin, April 2009.
8 See online at www.thelivesofspaces.com/ (accessed November 2011).
9 From the author's interview with Grainne Hassett, Dublin, April 2009.
10 Coop Himmelblau deliberately draw with their eyes closed, using their hands as seismographs to produce an initial drawing. Their first sketch is then repeatedly overlaid with further drawings and an architectural programme gradually introduced until the generating image disappears and the forms of the project emerge. The concept dictates that it is not the starting point that requires skill, but rather the interpretative process by the architect in deciding how to make use of apparently randomly generated lines and images to develop the final forms. If truly generated by chance, the authority of the design and the resultant architecture would be at

issue. Coop Himmelblau's method sought to achieve a result that was not accidental, although, in their case, the starting point was intended to be arbitrary. See Coop Himmelblau, *The Power of the City*, Vienna: Verlag der Georg Büchner, 1988.

11 Bart Lootsma, 'Ambidexterity and transgression', in Kristin Feireiss (ed.), op cit., 1994, pp. 19–27.

12 From the author's interview with Ben van Berkel and Caroline Bos, Amsterdam, September 2010.

13 Donald Schön, 'The reflective practitioner', quoted in David Nicol and Simon Pilling (eds), *Changing Architectural Education: Towards a New Professionalism*, London: E & FN Spon, 2000, pp. 15, 16 and viii.

14 Todd Gannon, op. cit., p. 14.

15 Ben van Berkel, from the author's interview with Ben van Berkel and Caroline Bos, Amsterdam, September 2010.

16 UN Studio research pamphlet for Akzo Nobel and in 'Out of the blue' article by Catherine Croft (*Ben van Berkel's 'Blue Period'* lecture at the AA) in *Building Design*, no. 1518, 1 February 2002, pp. 14–15.

17 *The Truman Show* (1998) is a feature film starring Jim Carrey and directed by Peter Weir in which the main character grows up in an entirely manmade stage-set town and his life is broadcast, unknown to him, as a TV soap.

18 Niall McLaughlin, a London-based architect, has made use of a similar iridescent material at his small housing project in the east end of London at Silvertown with artist Martin Richman. It is a subtle effect, gently diffused by a corrugated panel, but an intriguing one, given the rows of brick housing that surround the site. The effect of the material is similar to the sheen on a compact disc (CD), which is caused by light refracted among the profile of the deep rectangular troughs and valleys on the surface of the disc. Some flowers have developed a similar profile on their petals to attract birds that have an ability to see ultraviolet light.

19 From the author's interview with Ben van Berkel and Caroline Bos, Amsterdam, September 2010.

20 The term 'brandscapes' is thought to have been introduced by John Sherry in 1986 at the Conference of the Association for Consumer Research in Toronto, Canada. Subsequently used by Anna Klingmann, *Brandscapes: Architecture in the Experience Economy*, Cambridge, MA: MIT Press, 2010.

21 Andrea Pavoni, 'Erasing space from places, brandscapes, art and the (de)valorisation of the Olympic space' in *Lo Squaderno*, no. 18, December 2010, Explorations in Space and Society, pp. 9–13. Some of the text of this chapter first appeared in a paper by the author, Fiona McLachlan, for the AIC Conference in Zurich, June 2011.

11 Navigation, communication and language

1 Rolf Kuehni, 'Development of the idea of simple colours in the 16th and early 17th centuries' in *Color Research and Application*, Wiley, vol. 32, Issue 2, April 2007, pp. 92–99 notes that Aristotle considered that chromatic colours were generated by a mixture of black and white, with the fundamental colours being yellow, scarlet, purple, green and blue. Like Newton, Aristotle identified seven key colours and seems to share a desire to mimic the musical scale (p. 92).

2 Sir Isaac Newton's observation of the spectrum spilt through a prism was noted in 1790.

3 *Goethe's Colour Theory. Arranged and Edited by Rupprecht Matthei*, Herb Ach (transl.), London: Studio Vista, 1971. Matthei notes: 'Newton and Goethe followed totally different aims in their reason, while Newton attempted to analyze the nature of light, Goethe applied himself to the phenomenon of color', p. 6.

4 This is noted in the introduction by the translator in the later English edition of Johann Wolfgang von Goethe's *Theory of Colours*, English edition, London: Murray, 1840.

5 Ibid., p. 287.

6   'Hue' being the colour (e.g. red, blue), 'Value' being the lightness, and the 'Chroma' of a colour the intensity of its colouring. This is similar to the classification used in the RAL colour system.

7   Munsell's *Atlas* categorized into broad groupings:
    - *Hue*: Warm colours (red R, red-purple RP, yellow Y), Cool colours (green, blue, grey, white, black)
    - *Value*: from Very light 9/9.25 (reflectance 72–84 per cent), through Middle 5/6, (20–30 per cent) to Very dark 1/2 (1.5–2 per cent)
    - *Chroma*: Greys 0–0.5, Soft 1–3, Middle 4–9 and Strong 10 or more.

8   Rolf Kuehni and Andreas Schwarz, *Color Ordered: A Survey of Color Order Systems from Antiquity to the Present*, Oxford: Oxford University Press, 2008, gives a detailed history of the development of colour classification and navigation systems. An interactive compendium is available online at: www.colorsystem.com/ (accessed November 2011).

9   See online at www.sto.co.uk/92720_EN-StoColor_System-How_it_Works.htm (accessed November 2011). The StoColor System eschews colorimetric logic in favour of human perception of colour. It is based primarily on the colours yellow, orange, red, violet, blue and green. These six sections are mixed to form the 24 basic tones, and each basic tone is assigned five colour rows with a gradient of light to dark. Sto Render Guide, page 5, available at: www.sto.co.uk/66599_EN-brochures-StoRend-Brochure.htm (accessed November 2011).

10 See online at www.ncscolour.co.uk/ which gives a useful history of the development of the NCS system (accessed November 2011).

11 The distances between the individual colours are defined by the CIELAB-colour distance formula, which RAL notes is also embedded in DIN 6174.

12 For instance, a Dulux colour named Mystic Mauve 3 is coded as 54BB 41/237, which can be dissected into: Hue Family – BB (Blue through to Violet), LRV 41 (lowest being black, highest white, with most pastels having an LRV of between 75 and 83) and Chroma 237 (on a scale of 1,000 steps of saturation/intensity).

13 Taken from Dulux Trade literature *ICI 2000*.

14 Much of the early development work was carried out by local authority and government departments in the UK, in relation to the design and redecoration of schools and factories. The research, in consultation with architects, resulted in an extended range of 101 colours, which was adopted as the BS 2660: 1955 range, in turn adopted by the major paint manufacturers in the UK. Presently, the British Standard which provides a framework for colour coordination is BS 5252: 1976, of which the standard building and decoration range in BS 4800.

15 The artist Gerhard Richter also uses the term 'Atlas' to describe certain exploratory works. Gerhard Richter, Helmut Friedel and Städtische Galerie im Lenbachhaus München, *Atlas*, Cologne: Verlag der Buchhandlung Walther König, 1996.

16 Originally established in 1901 to develop standards for quality and sections in steel, the British Standards Institution has remained central in the development of international standardization.

17 ISO brought together the ISA (International Federation of the National Standardizing Associations), established in New York in 1926, and the UNSCC (United Nations Standards Coordinating Committee), established in 1944.

18 CEN are also responsible for the formal European Standards (EN), which are national standards that apply across all European Union member states through a network of national partner agencies, such as the German Deutsches Institut für Normung e.V. (DIN) and the Swedish Standards Institute (SIS). Each ISO standard which has been adopted by a member state is identified by a prefix. For instance:

- BS EN ISO 3668: 2001 *Paints and varnishes – Visual comparison of the colour of paints (ISO 3668: 1998)* – denotes that the BSI and CEN have adopted the International Standard
- DIN EN ISO 10545-16 *Ceramic tiles – Part 16: Determination of small colour differences (ISO 10545-16: 1999)* – denotes the German standard. All other member states have equivalents.

19 ICC available online at: www.color.org/aboutICC.xalter (accessed November 2011).
20 The values can also be represented on the Hunter L, a, b scale (where L = lightness–darkness, a = redness–greenness and b = yellowness–blueness).
21 Le Corbusier, *Polychromie architecturale: Farbenklaviaturen von 1931 und 1959/Color Keyboards from 1931 and 1959*, endnotes.
22 The RGB values used in computer monitors relate to the three 'additive primaries' of Red Green Blue, which together make white light. The CMYK values are based on the three 'subtractive primaries', which together make black. Printers add black (K) to vary the darkness.
23 Andrew Mead, 'Colour vision', in *Architects' Journal*, 11 October 2001, vol. 214, no. 13, pp. 34–43, Louisa Hutton quoted.
24 The ColorMunki is a small device that can 'piggyback' onto a computer monitor and calibrate it automatically. It can read colour from a surface and tease out colours from images on screen. It displays colour coordinates which define the exact CMYK reference.

12 Playing space: laws, rules and prescription

1 Johannes Itten, *The Art of Color: The Subjective Experience and Objective Rationale of Color*, New York: Reinhold Publications Co., 1961, p. 30.
2 Magdelena Drost, *Bauhaus, 1919–1933 Bauhaus-Archiv*, Koln: Taschen, 2002, p. 24. Johannes Itten began teaching at the Bauhaus in 1919. His 'Basic Course' was the foundation for teaching and was developed further by successive teachers, such as Paul Klee. Kandinsky's book *Concerning the Spiritual in Art* (1912) was highly influential in terms of his principles of primary colours and primary forms.
3 Bridget Riley, 'Colour for the painter' in Trevor Lamb and Janine Bourriau (eds), *Colour: Art and Science*, Cambridge: Cambridge University Press, 1995, p. 63.
4 Arthur Rüegg (ed.), *Polychromie architecturale: Les Claviers de Couleurs de Le Corbusier de 1931 et de 1959*, Basel: Birkhäuser, 1998, 2006, p. 99.
5 Izak Salomons (ed.), Aldo and Hanne van Eyck, *Built with Colour: The Netherlands Court of Audit by Aldo and Hanne van Eyck*, Rotterdam: 010 Publishers, 1999, p. 9.
6 Michael Lancaster, *Colourscape*, London: Academy Editions, 1996, p. 104. Also see Lancaster's Appendix, 'Colour guidelines'.
7 Ibid., p. 116.
8 *Amadeus* (1984), film script by Peter Shaffer, directed by Milos Forman.
9 The original French text (1839) does not have colour plates and there is a note to the effect that concerns about the sale price for the book restricted the publication. The later edition, in English (1859), does contain the colour plates.
10 Michel Eugène Chevreul, *The Laws of Contrast of Colour: and Their Application to the Arts of Painting, Decoration of Buildings, Mosaic Work, Tapestry and Carpet Weaving, Calico Printing, Dress, Paper Staining, Printing, Illumination, Landscape and Flower Gardening, etc.*, John Spanton (transl.), London: G. Routledge & Co., 1859 (English edition), French edition without colour plates, 1839, p. 9.

11   Ibid., p. 46. Personal experience suggests that this is still valid. In one project by the author, a rich, red cherrywood had been selected for custom-built cabinets. We had imagined this paired with warm blues, but the client insisted on yellow walls, which we felt did not work well with the woodwork.

12   Chevreul's laws of contrasting colours became widespread after the translation of the book into English in 1859. Texts on the subject by Hermann von Helmholtz, 1852 and, later, by Ogden Rood, *Modern Chromatics*, London: D. Appleton and Company, 1879, were all used by artists exploring ways to intensify colour. Philip Ball, *Bright Earth: The Invention of Colour*, Chicago and London: University of Chicago Press, 2001, gives a full account of these effects at pp. 189–207.

13   Michel Eugène Chevreul, op. cit., p. 176.

14   Josef Albers, *Interaction of Color*, New Haven and London: Yale University Press, 1963 (revised 2006). Albers joined the Bauhaus in 1923.

15   Department of Education and Science, Building Bulletin 9, *Colour in School Buildings*, 4th edition, London: HMSO, 1969, p. 6.

16   Ibid., p. 14.

17   Based on Munsell's definitions of hue (red, red-purple, blue, blue-green, green, etc.), value (lightness to darkness – paler *tint* to deeper *shade*) and chroma (strength or intensity).

18   On work surfaces, and floors, for instance, a high light reflectance value of a neutral colour will give good contrast to the activity. Dulux *Trade Colour Palette* categorizes each shade by Hue, Light Reflectance Value (LRV) and Chroma (e.g. 30 YR 53/188), making it simple to meet guidelines for light reflectance. This is of particular importance for people with visual impairment.

19   Software, such as NCS, RAL C1 Digital and ColorMunki (Pantone), offers such comparisons.

20   As part of the discourse on the meaning of 'surface' in architecture, the idea of self-imposed rules, particularly those relating to composition, has re-emerged, including the use of regulating lines or mathematical rules as parameters to constrain the placement of windows, the proportion of facades and their composition into a whole. As with colour, however, the composition of facades is now rarely taught in schools of architecture, and elevations tend to be considered as a by-product of the configuration of form and the relationship of internal to external space. 'Yet it is the facade that most directly communicates with the viewer; it is the frame through which we read the building and colour will make an immediate impression, before form or surface.' Colin Rowe and Robert Slutzky, 'Transparency: literal and phenomenal', 1976, reprinted in Colin Rowe, *The Mathematics of the Ideal Villa and Other Essays*, Cambridge, MA: MIT, pp. 2–16 and 159–83.

21   Lucila Geymonat de Destefani and T.W. Allen Whitfield, 'Esthetic decision-making: how do people select colours for real settings?' in *Color Research and Application*, vol. 33, no. 1, February 2008.

22   Johannes Itten and Faber Birren, *The Elements of Color: A Treatise on the Color System of Johannes Itten, Based on his Book 'The Art of Color', 1961*, London: John Wiley & Sons, 1970, p. 24. Le Corbusier also makes reference to the influence of personality on colour choice in the introductory notes to the 1959 *Salubra* colour collection: 'within these 400 combinations, the user will find the color equilibrium which will conform with his own nature'.

23   Josef Albers, 1963, op. cit., p. 17.

24   For Sigmund Freud, this duality became that of the 'ego' (conscious and rational) and the 'id' (the hotbed of emotion and crude desires). Immanuel Kant considered moral instincts to be ruled by rationality, but recently advances in neurological studies have enabled brain patterns to be traced to demonstrate that morality is much more often instinctive and immediate. The emotional brain kicks in first and, only afterwards, do we rationalize.

Jonah Lehrer, *The Decisive Moment*, Edinburgh: Canongate Press, 2009, p. 18. As Lehrer notes: 'People often come up with persuasive reasons to justify their moral intuition', p. 167.

25 John Gage, *Colour and Meaning: Art Science and Symbolism*, Berkeley, CA: University of California Press, London: Thames & Hudson, 2000, p. 35. In discussing gender issues in relation to colour, Gage makes reference to the nineteenth-century theorist Charles Blanc, who considered drawing to be the masculine form of art and colour as the feminine, and Gage suggests 'for that reason colour could only be of secondary importance'. He continues to note that, while Philipp Otto Runge believed yellow and red to evoke the masculine passions and blue and violet the feminine, more recently opinion has reversed, with blue seen as masculine. Gage concludes the section 'perhaps the most interesting area for feminists to explore is, indeed, the recurrent assumptions that a feeling for colour is itself a peculiarly female province'.

26 Ludwig Wittgenstein, *Remarks on Colour*, no. 22, I–24, 1977, p. 5e.

27 John Gage, *Colour and Culture*, Berkeley, CA: University of California Press, London: Thames & Hudson, 1999, p. 188.

– BIBLIOGRAPHY –

Abrahams, T., 'Barking Central', *Blueprint*, 291, June 2010, pp. 36–42

Adjaye, D., *Making Public Buildings*, Allison, P. (ed.), London: Thames & Hudson, 2006

Albers, J., *Interaction of Color*, New Haven, CT and London: Yale University Press, 1963 (revised 2006)

Allen, I.; Dawson, S. 'A bright future [Jubilee School, London]', *The Architects' Journal*, vol. 217, no. 17, 1 May 2003, pp. 28–37

Allen, J., *Designer's Guide to Color*, London: Angus & Robertson Publishers, 1986

Anter, K.F., 'Forming spaces with colour and light: trends in architectural practice and Swedish colour research', *Colour: Design & Creativity*, vol. 2, 2008, pp. 2, 1–10

Architects' Journal, vol. 229, no. 6, 19 February 2009, pp. 26–37

Architectural Record, vol. 196, no. 5, May 2008, pp. 206–13

Asensio, P.; Cuito, A., *Legorreta + Legorreta*, New York: Kempen teNeues Publishing Group, 2002

Attenbury, P.; Wainwright, C., *Pugin: A Gothic Passion*, New Haven, CT, and London: Yale University Press in association with the Victoria and Albert Museum, 1994

A + U: architecture and urbanism, October 2002, no. 385, pp. 76–81

A + U: architecture and urbanism, June 2004, no. 405, pp. 76–83, 88–95

A + U: architecture and urbanism, December 2006, no. 435, pp. 80–87

A + U: architecture and urbanism, May 2008, no. 452, pp. 38–47

Bachmann, U. *Farbe und Licht/Colour and Light*, Zurich: Verlag Niggi AG (book and DVD-rom), 2011

Bakacsy, J.; Munch, A.V.; Sommer, A-L., *Architecture, Language, Critique: Around Paul Engelmann*, Amsterdam: Editions Rodopi BV, 2000

Ball, P., *Bright Earth: The Invention of Colour*, Chicago and London: University of Chicago Press, 2001

Banachek, *Psychological Subtleties 2*, Houston: Magic Inspirations, 2007

Batchelor, D., *Chromophobia*, London: Reaktion Books, 2000

Baumeister, vol. 102, no. 3, March 2005, pp. 20–21

Baus, U., 'La brillance de la couleur Erich Wiesner et Otto Steidle', *Archithese – Niederteufen*, Switzerland: Niggli Ltd, 2003, Part 5, pp. 76–77

Berkel, B. van, *Mobile Forces/Mobile Kräfte*, Feireiss, K. (ed.), Berlin: Ernst & Sohn, 1994

Berkel, B. van; Bos, C., *Delinquent Visionaries*, Amsterdam: Ben van Berkel, 1993

Betsky, A. (ed.); Berkel, B. van; Bos, C., *UN Studio FOLD*, Rotterdam: NAi Publishers, 2002

Birren, F., *Principles of Color*, New York: Van Nostrand Reinhold, 1969

Birren, F., *Color and Human Response*, New York: Van Nostrand Reinhold, 1978

Birren, F., *Light, Color and Environment: A Discussion of the Biological and Psychological Effects of Color, With Historical Data and Detailed Recommendations for the Use of Color in the Environment*, New York and London: Van Nostrand Reinhold, 1982

Birren, F., *Light, Color and Environment* (revised edition), New York: Van Nostrand Reinhold, 1982

Borden, I. (ed.), *Strangely Familiar*, London: Routledge, 1996

Borden, I., *Manual: The Architecture and Office of Alford Hall Monaghan Morris*, Basel: Birkhäuser, 2003

Boulding, K., *The Image: Knowledge in Life and Society*, Michigan: University of Michigan Press, 1956 and 1961

Braham, W.W., *Modern Color/Modern Architecture*, Farnham, Hampshire: Ashgate, 2002

Brenne, W., *Bruno Taut Meister des farbigen Bauens in Berlin*, Berlin: Braun 2008

Bürkle, J.C.; Landert, M. (eds); Gigon, A.; Guyer, M., *Gigon/Guyer: Works and Projects 1989–2000*, Zurich: Berlag Niggli AG, 2000

Caivano, J.L., 'Research on color in architecture and environmental design, current developments, and possible future', *Color Research & Application*, vol. 31, no. 4, August 2006

Campbell, H., 'Lesson plan; Architects: O'Donnell + Tuomey', *RIBA Journal*, vol. 105, no. 12, December 1998, pp. 28–35

Cameron, N., *Graven Images: Design in a Cold Climate*, University of Michigan, Frame Publishers, 2002

Canter, D., *The Psychology of Place*, London: Architectural Press, 1977

Caruso, A.; St John, P.; Coombs, M., 'A sense of the organic', *The Architects' Journal*, vol. 202, 13 July 1995, pp. 27–33

Caruso St John Architects, *Knitting Weaving Wrapping Pressing*, Basel: Birkhäuser, 2002

Caruso St John Architects, *The New Art Gallery, Walsall*, London: Batsford, 2002

Castello, L., *Rethinking the Meaning of Place: Conceiving Place in Architecture-Urbanism*, Farnham, Hampshire: Ashgate, 2010

Chevreul, M.E., *The Principles of Harmony and Contrast of Colours*, 2nd edition, London: Longman, Brown, Green and Longman, 1855

Chevreul, M.E., *The Laws of Contrast of Colour: and Their Application to the Arts of Painting, Decoration of Buildings, Mosaic Work, Tapestry and Carpet Weaving, Calico Printing, Dress, Paper Staining, Printing, Illumination, Landscape and Flower Gardening, etc.*, Spanton, J. (trans.), London: G. Routledge & Co., 1859 (English edition) French edition, without colour plates, 1839

Chipp, H.B.; Selz, P.H., *Theories of Modern Art: A Source Book by Artists and Critics*, Berkeley, CA: University of California Press, 1968

Chmielewska, E., 'Implaced communication: wayfinding and informational environments', PhD thesis, Department of Art History and Communication Studies, Montreal: McGill University, January 2001, available at: http://digitool.library.mcgill.ca/R/?func=dbin-jump-full&object_id=36891&local_base=GEN01-MCG02 (accessed November 2011)

Cohen, J.-L., *Le Corbusier, 1887–1965: The Lyricism of Architecture in the Machine Age*, Berlin: Taschen, 2004

Cohen, J.-L.; Moeller, G.M. Jr. (eds), *Liquid Stone: New Architecture in Concrete*, Basel: Birkhäuser, 2006

Colomina, B.; Bloomer, J., *Sexuality and Space*, New York: Princeton Architectural Press, 1992

Cook, P., 'To understand new architecture fully, nothing lives up to the experience of an actual visit', *Architectural Review*, March 2007, vol. 221, no. 1321, p. 34

Coop Himmelblau, *The Power of the City*, Vienna: Verlag der Georg Büchner, 1988

Cornelissen, H. (ed.), *Dwelling as a Figure of Thought*, Amsterdam: Sun Publishers, 2005

Cowling, E., *Interpreting Matisse Picasso*, London: Tate Publishing, 2002

Croft, C., 'Out of the blue' (Ben van Berkel's 'Blue Period' lecture at the AA), *Building Design*, no. 1518, 1 February 2002, pp. 14–15

Daidalos, guest editor Gerhard Auer, no. 51, 15 March 1994, p. 45

Davey, P., 'True colours', *Architectural Review*, vol. 204, no. 1221, November 1998, pp. 34–35

Department of Education and Science Building Bulletin 9, *Colour in School Buildings*, 4th edition, London: HMSO, 1969

Destefani, G. de; Whitfield, A., 'Esthetic decision-making: how do people select colours for real settings?', *Color Research and Application*, vol. 33, no. 1, February 2008

Dixon, A.G., *Howard Hodgkin*, London: Thames & Hudson, 1994

Doherty, G. (ed.), *New Geographies 3: Urbanisms of Color*, Cambridge, MA: Harvard University Press, 2010

Droste, M., *Bauhaus, 1919–1933*, Bauhaus-Archiv, Koln: Taschen, 2002

Duttman, M.; Schmuck, F.; Uhl, J., *Color in Townscape*, London: The Architectural Press, 1981

Eco, U., 'How culture conditions the colours we see', in Blonsky, M., *On Signs*, Baltimore, MD: JHU Press, 1985

Edinburgh World Heritage, *Historic Home Guide: External Paintwork*. Available at: www.ewht.org.uk/uploads/downloads/EWH%20External%20Paintwork%20Guide.pdf (accessed November 2011)

Elger, D., *Donald Judd: Colorist*, Stuttgart: Edition Cantz, 2000

Faulkner, W., *Architecture and Color*, Chichester: Wiley Interscience, 1972

Finlay, V., *Colour: Travels Through the Paintbox*, London: Hodder & Stoughton, 2002

Foreign Office Architects, Institute of Contemporary Arts, London, England, *Phylogenesis: Foa's Ark*, Barcelona: Actar, 2008

Forster, K.; Sauerbruch, M.; Hutton, L.; Mostafavi, M., *WYSIWYG*, London: Architectural Association, 1999

Frampton, K., 'Towards a critical regionalism: six points for an architecture of resistance', in Foster, H. (ed.), *The Anti-Aesthetic*, Seattle: Bay Press, 1983

Frampton, K., *Steven Holl, Architect*, Milan: Electa Architecture, distributed by Phaidon Press, 2003

Futagawa, Y., *Ricardo Legorreta*, Tokyo: A.D.A Edita, 2000

GA document, no. 60, November 1999, pp. 66–75

GA document, no. 67, October 2001, pp. 36–43

GA document, no. 96, May 2007, pp. 90–107

GA document, no. 97, June 2007, pp. 54–57

Gage, J., *Colour and Culture*, London: University of California Press/Thames & Hudson, 1999

Gage, J., *Colour and Meaning: Art, Science and Symbolism*, London: University of California Press/Thames & Hudson, 2000

Gannon, T., *UN Studio: Erasmus Bridge, Rotterdam, The Netherlands*, New York: Princeton Architectural Press, 2004

Gannon, T. (ed.), *Steven Holl Simmons Hall*, New York: Princeton Architectural Press, 2004

Gans, D., *Le Corbusier Guide*, New York: Princeton Architectural Press, 2006

Garfield, S., *Mauve*, London: Faber & Faber, 2000

Gatz, K., *Colour and Architecture*, London: Batsford, 1967

Gibson, R., 'Above par for the course: O'Donnell + Tuomey', *The Architects' Journal*, vol. 201, no. 20, 25 May 1995, pp. 25–34

Giedion, S., *Space, Time and Architecture: The Growth of a New Tradition*, Cambridge, MA: Harvard University Press, 1st edition 1941, 2008

Gigon, A.; Guyer, M.; Yoshida, K., *Gigon/Guyer: Matter, Colour, Light and Space*, Tokyo: A+U Publishing Co., 2006

Gloag, H.L., 'Building Research Station', in *Factory Building Studies No 8: Colouring in Factories*, London: HMSO, 1961

Goethe, J.W. von (1749–1832), *Die Farbenlehre* 1810, *Theory of Colours*, English edition London: Murray, 1840

Goethe, J.W. von; Matthei, R., *Goethe's Colour Theory*, arranged and edited by R. Matthei, H. Aach (transl.), London: Studio Vista, 1971

Gomez-Perez, A.; Pelletier, L., *Architectural Representation and the Perspective Hinge*, Cambridge, MA: MIT Press, 1997, 2000

Gregory, R., 'AHMM: Westminster City Academy, London, UK', *Architectural Review*, vol. 217, no. 1298, April 2005, p. 66

Hadid, Z., *Planetary Architecture ll*, folio, London: Architectural Association, 1983

Hamm, O. (ed.), *Steidle + Partner Wohnquartier Freischutstraße, München*, Stuttgart: Axel Menges, 2003

Harbison, R., 'Much still to learn about using colour' [exhibition review], *Architects' Journal*, vol. 197, no. 12, 24 March 1993, p. 52

Hardy, A., *Colour in Architecture*, London: L. Hill, 1967

Hawkins, D., 'Colour theory: a critical perspective from Besant and Leadbetter to Sauerbruch Hutton Architects', dissertation submitted in partial fulfilment of the requirements for an Edinburgh College of Art Diploma in Architecture, 2004

Hay, D.R., *The Laws of Harmonious Colouring: Adapted to Interior Decorations, Manufacturers and Other Useful Purposes*, London: W.S. Orr, Edinburgh: W. & R. Chambers, 1828

Heer, J. de, *The Architectonic Colour: Polychromy in the Purist Architecture of Le Corbusier*, Rotterdam: 010 Publishers, 2009

Hodgson, P.H.; Toyka, R. (eds), *The Architect, the Cook and Good Taste*, London: Birkhäuser, 2007

Holl, S., *Anchoring*, New York: Princeton Architectural Press, 1989, 1991

Holl, S., *Intertwining*, New York: Princeton Architectural Press, 1996

Holl, S., *Parallax*, New York: Princeton Architectural Press, 2000

Holl, S.; Pallasma, J.; Gomez-Perez, A. *Questions of Perception: Phenomenology of Architecture*, San Francisco: William Stout, 2006

Hubertus, A.; Moos, S. von., *El Croquis*, no. 143, 2009, pp. 12–13

Hyman, J., *The Objective Eye: Colour, Form and Reality in the Theory of Art*, Chicago and London: University of Chicago Press, 2006

Imperial Chemical Industries, *Landmarks of the Plastics Industry*, London: Kynoch Press, 1962

Inberg, H., 'Sampling and re-mixing an architecture of resistance', *Canadian Architect*, vol. 44, issue 9, September 1999, pp. 26–33

Itten, J., *The Art of Color: The Subjective Experience and Objective Rationale of Color*, New York: Reinhold Publications Co., 1961

Itten, J.; Birren, F., *The Elements of Color: A Treatise on the Color System of Johannes Itten, Based on his Book 'The Art of Color'*, London: John Wiley & Sons, 1970

Jarman, D., *Chroma*, Vintage: London, 1995

Jones, O., *The Grammar of Ornament*, London: Day & Son Ltd, 1856

Kandinsky, W., *Concerning the Spiritual in Art*, translated from the German and with an introduction by M.T.H. Sadler, New York: Dover Publications, London: Constable, 1977, reprinted Las Vegas: IAP, 2009 (original English edition *The Art of Spiritual Harmony*, London: Constable, 1914, translation of *Über das Geistige in der Kunst*, München: Piper, 1912)

Kant, I., *Critique of Judgement*, originally published 1914, New York: Cosimo, 2007

Kay, P.; Berlin, B., *Basic Color Terms: Their Universality and Evolution*, Berkeley, CA: University of California Press, 1969, 1991

Kemp, M., *The Science of Art: Optical Themes in Western Art from Brunelleschi to Seurat*, New Haven and London: Yale University Press, 1990

Kirsch, K., *The Weissenhofseidlung Experimental Housing for the Deutscher Werkbund, Stuttgart 1927*, New York: Rizzoli, 1989

Kisby, S., 'Bruno Taut: architecture and colour' (essay from the Welsh School of Architecture online at: www.kisbee.co.uk/sarc/taut/taut.htm (accessed 29 February 2012)

Klingmann, A., *Brandscapes: Architecture in the Experience Economy*, Cambridge, MA: MIT Press, 2010

Koolhaas, R., *Colours*, Basel: Birkhäuser, 2001

Koolhaas, R.; Mau, B.; Werlemann, H., Office for Metropolitan Architecture, *Small, Medium, Large, Extra-Large (S,M,L,XL)*, New York: Monacelli Press, 1995

Kossak, F. (ed.), *Otto Steidle Structures for Living*, Zurich: Artemis, 1994

Kudielka, R., *Robert Kudielka on Bridget Riley*, London: Ridinghouse, 2005

Kudielka, R.; Shone, R., *Bridget Riley: Dialogues on Art*, London: Thames & Hudson, 2003

Kuehni, R.G., 'Development of the idea of simple colours in the 16th and early 17th centuries', *Color Research and Application*, vol. 32, issue 2, April 2007, John Wiley & Sons, pp. 92–99

Kuehni, R.G.; Schwarz, A., *Color Ordered: A Survey of Color Order Systems from Antiquity to the Present*, Oxford: Oxford University Press, 2008

Lamb, T.; Bourriau, J. (eds), *Colour: Art and Science*, Cambridge: Cambridge University Press, 1995

Lancaster, M., *Colourscape*, London: Academy Editions, 1996

Leatherbarrow, D.; Mostafavi, M., *On Weathering: The Life of Buildings in Time*, 2nd edition, Cambridge, MA: MIT, 1993/1997

Leatherbarrow, D.; Mostafavi, M., *Surface Architecture*, Cambridge, MA: MIT, 2002

Le Corbusier, *The Modulor: A Harmonious Measure to the Human Scale Universally Applicable to Architecture and Mechanics*, originally published in 1950, London: Faber & Faber, 1951

Legorreta, R.; Legorreta, V., *Legorreta plus Legorreta*, New York: Rizolli International Publications, 2004

Lehrer, J., *The Decisive Moment*, Edinburgh: Canongate Press, 2009

Lemon, L.T.; Reis, M.J. (transl.), *Russsian Formalist Criticism: Four Essays*, Omaha: University of Nebraska Press, 1965

Lenclos, J.-P.; Lenclos, D., *Colors of the World: The Geography of Color*, English edition, 2004, New York: W.W. Norton & Company, original French edition 1999, Paris: Groupe Moniteur

Linton, H., *Color in Architecture: Design Methods for Buildings, Interiors and Urban Spaces*, New York and London: McGraw Hill, 1999

Lissitzky-Kuppers, S., *El Lissitzky: Life, Letters, Texts*, London: Thames & Hudson, 1980

Loeb, C.; Loeb, A. (eds), *Theo van Doesburg On European Architecture: Complete Essays from Het Bouwbedrift 1924–1931*, Basel: Birkhäuser Verlag, 1990

Lootsma, B., 'Diagrams in costumes', *A+U: Architecture and Urbanism*, no. 3(342), March 1999, pp. 98–103

Lynch, K., *The Image of the City*, Cambridge, MA: MIT, 1960

McCown, J.; Ojeda, O.R.; Warchol, P., *Colors: Architecture in Detail*, Minneapolis, MN: Rockport Publishers, 2006

McLachlan, F., 'Dancing windows', *ARQ 2006*, vol. 10, pp. 191–200

McLachlan, F.; Coyne, R., 'The accidental move: accident and authority in design discourse', *Design Studies*, no. 22, 2001

McLean-Ferris, L., Nottingham Contemporary artreview.com, 17 November 2009, available at: www.artreview.com/forum/topic/show?id=1474022%3ATopic%3A929663 (accessed November 2011)

Manke, F.H., *Color, Environment and Human Response: An Interdisciplinary Understanding of Colour and its Use as a Beneficial Element in the Design of the Architectural Environment*, London: John Wiley & Sons, 1996

Mead, A., 'Colour vision', *Architects' Journal*, vol. 214, no. 13, 11 October 2001, pp. 34–43

Meade, M., *Special Issue: Leisure, Architectural Review*, vol. 198, no. 1186, December 1995, pp. 4–5, 27–77

Meerwein, G.; Rodeck B.; Mahnke, F., *Color: Communication in Architectural Space*, Basel: Birkhäuser, 2007

Miller, M.C., *Color for Interior Architecture*, New York and London: John Wiley & Sons, 1997

Moore, R., 'A pebble on water', *New Art Gallery Walsall*, London, UK: Batsford, 2002

Morgan, W., *The Elements of Structure*, London: Pitmann, 1971

Morle, J., *Architecture in Perspective: Construction, Representation, Design and Colour*, London: Batsford, 1994

Mullin, S., 'A class of its own' [Adelaide Wharf, London], *Architectural Review*, vol. 223, no. 1341, November 2008, pp. 86–91

Newman, J.O., *Adolf Loos, Spoken into the Void: Collected Essays 1897–1900*, J.H. Smith (transl.), includes Loos, A., *The Principle of Cladding* (Das Prinzip der Bekleidung), Cambridge, MA: MIT Press, 1982

Nicol, D.; Pilling, S. (eds), *Changing Architectural Education: Towards a New Professionalism*, London: E & FN Spon, 2000

Norberg-Schultz, C., *Genius Loci: Towards a Phenomenology of Architecture*, London: Academy Editions, 1980

O'Connor, Z., 'Colour harmony revisited', *Color Research & Application*, vol. 35, 2010, pp. 267–73, doi: 10.1002/col.20578

O'Donnell + Tuomey: Casa gigante dormido, Killiney, Sleeping Giant House, Killiney (Ireland), *AV monographs*, no. 127, September–October 2007, pp. 40–43

Ostwald, W., *The Colour Primer*, Birren, F. (ed.), New York & London: Van Nostrand Reinhold, 1969

Ostwald, W.; Scott Taylor, J., *The Ostwald Colour Album*, London and New York: Winsor & Newton, 1934

Ozenfant, A., 'Colour: experiments, rules, facts', *Architectural Review*, vol. 81, April 1937

Pallasmaa, J., *The Eyes of the Skin: Architecture and the Senses*, London: Academy Editions, 1996

Passini, R., *Wayfinding in Architecture*, New York: Van Rostrand Reinhold, 1984

Pauly, D., *Barragán: Space and Shadow, Walls and Colour*, Basel: Birkhäuser, 2002

Pavey, D. (ed.), *Color*, Los Angeles: The Knapp Press, 1980

Pavoni, A., 'Erasing space from places, brandscapes, art and the (de)valorisation of the Olympic space', *Lo Squaderno*, no. 18, December 2010

Penoyre and Prasad, *Tranformations: The Architecture of Penoyre and Prasad*, London: Black Dog Publishing, 2007

Porter, T., *Colour Outside*, London: The Architectural Press, 1982

Porter, T., *Will Alsop: The Noise*, London: Routledge, 2011

Porter, T.; Mikellides, B. (eds), *Colour For Architecture*, London: Studio Vista; New York: Van Nostrand Reinhold, 1976

Porter, T.; Mikellides, B. (eds), *Colour for Architecture Today*, London and New York: Routledge, 2009

Powell, K., 'Adding colour to the city', *Architectural Review*, vol. 219, no. 1308, February 2006, pp. 68–71

Rasmussen, S.E., *Experiencing Architecture*, Cambridge, MA: MIT Press, 1959

Rattenbury, K., '*The sea, the sea*; Architects: O'Donnell + Tuomey', *RIBA Journal*, vol. 111, no. 6, June 2004, pp. 36–42

Rattray, C.; Hutton, G., 'Concepts and material associations in the work of Gigon/Guyer', *Architectural Research Quarterly (ARQ)*, vol. 4, no. 1, 2000, p. 20

Rehsteiner, J.; Sibillano, L.; Wettstein, S., *Farbraum Stadt: Box ZRH Das Buch Eine Untersuchung und ein Arbeitswerkzeug zur Farbe in der Stadt*, Zurich: Haus der Farbe, CRB, Kontrast, 2010

Rendell, J., *Art and Architecture: A Place Between*, London: I.B. Tauris, 2006

Richardson, V., 'Out of place', *RIBA Journal*, vol. 108, no. 11, November 2001, pp. 16–18

Richter, G.; Friedel, H.; Städtische Galerie im Lenbachhaus München, *Atlas*, Köln: Verlag der Buchhandlung Walther König, 1996

Richter, G., *Text 1961–2007*, Köln: Verlag der Buchhandlung Walther König, 2008

Riley II, C.A., *Color Codes: Modern Theories of Color in Philosophy, Painting and Architecture, Literature, Music, and Psychology*, Hanover, NH: University Press of New England, 1995

Risselada, M. (ed.), *Raumplan versus Plan Libre Adolf Loos and Le Corbusier 1919–1930*, New York: Rizzoli, 1988

Rood, O., *Modern Chromatics*, London: D. Appleton and Company, 1860

Rossi, A.A., *Scientific Autobiography*, Cambridge, MA: MIT Press, 1981

Rottenschlager, A., 'The sound of the cyborg', *The Red Bulletin*, March, 2011, www.neilharbisson.com/ (accessed November 2011)

Rowe, C., *The Mathematics of the Ideal Villa and Other Essays*, Cambridge, MA: MIT Press, 1976

Rüegg, A. (ed.), *Polychromie Architecturale: Les Claviers de Couleurs de Le Corbusier de 1931 et de 1959*, Basel: Birkhäuser, 1998, 2006

Runge, P.O. (Mahler), *Farben-Kugel oder Construction des Verhältnisses aller Mischungen der Farben zu einander, und ihrer vollständigen Affinität, mit angehängtem Versuch einer Ableitung der Harmonie in den Zusammenstellungen der Farben*, Hamburg, 1810

Russell, B., 'Color: the architectural search for joy', *House and Garden*, vol. 90, March 1976

Salomons, I. (ed.); van Eyck, A.; van Eyck, H., *Built with Colour: The Netherlands Court of Audit by Aldo and Hanne van Eyck*, Rotterdam: 010 Publishers, 1999

Sauerbruch Hutton, *Projekte 1990–1996 Architecture in the New Landscape*, Basel: Birkhäuser, 1996

Sauerbruch Hutton, *WYSIWYG*, London: Architectural Association, 1999

Sauerbruch Hutton, *Sauerbruch Hutton Archive*, Baden: Lars Müller Publishers, 2006

Sauerbruch, M.; Hutton, L.; Betsky, A., 'Sauerbruch Hutton Architects 1997–2003', *Croquis*, no. 114[I], 2003, entire issue (132 pp.)

Scanlon, E., 'Community Centre, Dublin; Architects: Hassett Ducatez Architects', *A10*, no. 21, May/June 2008, pp. 39–40

Schiess, A.; Wechsler, M.; Wirz, H.; Herzog & de Meuron; Gigon & Guyer, *Adrian Schiess: Farbräume: Zusammenarbeit mit den Architekten Herzog & de Meuron und Gigon/Guyer 1993–2003* (*Colourspaces: Collaboration with the Architects Herzog & de Meuron and Gigon/Guyer 1993–2003*), Luzern: Quart, 2004

Semper, G., *Style in the Technical and Techtonic Arts: Der Stijl*, Los Angeles: Getty Research Institute, 2004

Silvestrini, N., *Idee Farbe*, Zurich: Baumann & Stromer, 1994

Simpson, O. von, *The Gothic Cathedral*, New York: Pantheon, 1956

Slessor, C., 'Urban geometry: housing, Dalston, London', *Architectural Review*, vol. 206, no. 1233, November 1999, pp. 58–60

Solon, L.V., *Polychromy: Architectural and Structural Theory and Practice*, New York: The Architectural Record, 1924

Sottsass, E.; Radice, B. (ed.), *Notes on Colour*, Italy: Abet Edizioni, 1993

Spetz, R., 'Luxuriuos', *Arquitecturas de Autor*, Issue 13, 2000, pp. 4–7

Stansfield, J.; Allan Whitfield, T.W., 'Can future colour trends be predicted on the basis of past colour trends? An empirical investigation', *Color Research & Application*, vol. 30, Issue 3, pp. 235–42, available online at: http://onlinelibrary.wiley.com/doi/10.1002/col.20110/abstract (accessed November 2011)

Stockton, J., *Designer's Guide to Color*, London: Angus & Robertson Publishers, 1984

Stockton, J., *Designer's Guide to Color 2*, London: Angus & Robertson Publishers, 1985

Talbert, R., *Paint Technology Handbook*, London: CRC Press, 2008

Taut, B., *Ein Wohnhaus*, Stuttgart: Franckh'sche Verlagsbuchhandlung W. Keller & Co., 1927

Toy, M. (ed.), *Colour in Architecture. Architectural Design*, vol. 66, issues 3–4, London: Academy Editions, 1996

Tuomey, J., *Architecture, Craft and Culture*, Kinsale Ireland: Gandon Editions, 2004

Tuomey, J.; O'Donnell, S.; McGuire, B., *The Irish Pavilion*, Gandon Editions Works 8, 1992

Turnbull, A. 'Where things meet … light, colour, corner, edge', *Architects' Journal*, vol. 204, no. 16, 31 October 1996, pp. 34–35

Venturi, R., *Complexity and Contradiction in Architecture*, London: The Architectural Press, 1977

Walpamur Co. Ltd, *Colour in Buildings*, London: Lonsdale-Hands Associates, 1959

Wang, W., 'The variegated minimal', *El Croquis*, vol. 102, 2000, p. 27

Watkin, D., *Morality and Architecture*, Oxford: Clarendon Press, 1977

Werner, A.G.; Syme, P., *Werner's Nomenclature of Colours: With Additions, Arranged So as to Render It Highly Useful to the Arts and Sciences. Annexed to which are Examples Selected from Well-Known Objects in the Animal, Vegetable, and Mineral Kingdoms*, Edinburgh: W. Blackwood, 1814

Westland, S.; Laycock, K.; Cheung, V.; Henry, P.; Mahyar, F., 'Colour harmony', *Colour: Design & Creativity*, vol. 1, no. 1, 2007, pp. 1–15

Weston, R., *Materials, Form and Architecture*, New Haven, CT: Yale University Press, 2003

Whittle, S., 'Uncertain harmonies', *Colour: Design & Creativity*, vol. 1, no. 1: 10, 2007, pp. 1–4

Whyte, I.B., *Bruno Taut and the Architecture of Activism*, Cambridge: Cambridge University Press, 1982

Whyte, I.B. 'The Sublime: an introduction', in R. Hoffmann and I. Boyd Whyte (eds), *Beyond the Finite: The Sublime in Art and Science*, New York: Oxford University Press, 2011

Wigley, M., *White Walls, Designer Dresses: The Fashioning of Modern Architecture*, Cambridge, MA: MIT Press 1995

Wilkinson, J.G., *On Colour: And on the Necessity for a General Diffusion of Taste Among All Classes*, London: J. Murray, 1858

Williams, T.; Tsien, B., *O'Donnell + Tuomey: Selected Works*, New York: Princeton Architectural Press, 2007

Wilson, P.; Bolles, J., *Western Objects, Eastern Fields: Recent Projects by the Architekturbüro Bolles Wilson*, London: Architectural Association, 1989

Wittgenstein, L., *Bemerkungen über die Farben*, Berkeley, CA: University of California Press, 1977, English edition Oxford: Blackwell Publishing, 1979

Wittgenstein, L., *Remarks on Colour* (*Bemerkungen über die Farben*), G.E.M. Anscombe, L.L. McAlister, M. Schattle (transl.), Oxford: Blackwell Publishing, 1979

Wittkower, R.; Wittkower, M.; Roseborough Collins, G., *Gothic Versus Classic: Architectural Projects in Seventeenth Century Italy*, London: Thames & Hudson, 1974

Zelanski, P.; Fisher, M.P., *Colour*, London: Prentice Hall, 1989, 3rd edition, 1999

Zelner, P. (curator), *Sign as Surface*, exhibition catalogue, Architecture and Design Project Series, New York: Artists Spaces, 2004

IMAGE CREDITS

Every effort has been made to contact and acknowledge copyright owners, but the author and publisher would be pleased to have any errors or omissions brought to their attention so that corrections may be published at a later printing. All images by the author unless otherwise stated. Page numbers in bold.

xii	Image courtesy of Blue Tongue Entertainment
3	© James Turrell, Photo: Florian Holzherr
5	Drawing: E & F McLachlan Architects
6	Photo: Rachel Travers
8	Photo: E & F McLachlan Architects
12	© Allford Hall Monaghan Morris/Tim Soar
13	Photo: Margherita Spiluttini
15	Ben van Berkel/UN Studio, Photo: © Christian Richters
16	(bottom) Image: Bolles + Wilson
16	(top) Photo: Bolles + Wilson
22	Photos: Iain Stewart
25	Photo: © John Searle
28	Image: Victoria and Albert Museum
31	Photo: Boyer/Roger Viollet/Getty Images
36	Image: Jill Stansfield and T.W. Allan Whitfield
42	© Thomas Demand, Photo: Annette Kisling, Berlin
48	Photo: © Denis Gilbert/VIEW
49	Photo: Ewen McLachlan
51	Image: Yorck project Wikipedia Commons
52	Diagram by Johannes Itten, redrawn by the author
54	(top) Photo: © Ros Kavanagh/VIEW
56	Photo: © John Searle
59	Image: O'Donnell + Tuomey Architects
55	Photo: © John Searle
60	Photo: © Denis Gilbert/VIEW
63	© Will Alsop
67	© Judd Foundation. Licensed by VAGA, New York/DACS, London, 2011, Photo © Tate, London 2011
68	© Bridget Riley 2011. All rights reserved, courtesy Karsten Schubert, London
72–3	Photo: Reinhard Görner, goerner-foto.de
77–8	Photo: Reinhard Görner, goerner-foto.de
79	Photo: Ewen McLachlan
82	Photo: Paul Warchol Photography Inc.
84	Photo: Ken McCown
85	Image: Steven Holl Architects
86	Photos: Mik Gruber (ZHdK). Copyright Florian Bachmann/Ulrich Bachmann. First published in Ulrich Bachmann, *Farbe und Licht/Colour and Light*, Niggli Publishers, Switzerland, 2011
88	(top) Photo: Paul Warchol Photography Inc.
88	(bottom) Photo: Paul Warchol Photography Inc.
90	Photo: Iwan Baan
91	Photos: Doreen Balabanoff
92	Images: Leonhard Oberascher. First published in AIC 2011 Conference paper, Zurich
93	Image: FAP 25 (recto and verso) in *Farbraum Stadt. Box ZRH*, Jürg Rehsteiner, Lino Sibillano, Stefanie Wettstein (eds), Haus der Farbe Zurich/Berlin, Zurich 2010. Painter of the portrait: Simone Egger Foto on the verso: AnneMarie Neser © Haus der Farbe
94	Photo: Courtesy of Legorreta + Legorreta, photographer: Lourdes Legorreta
96	Photo: Achim Behn, Graz
95	Photo: E & F McLachlan
98–9	Image: Guy Nordenson, courtesy Steven Holl Architects
106	Photo: Serge Demailly
118	Photo: © Allford Hall Monaghan Morris/Tim Soar
120	Photo: Morag Myerscough
121	Photo: © Allford Hall Monaghan Morris/Tim Soar
122	Photo: Deirdre & Eleanor Evans

Image credits

124	(top) Photo: © Allford Hall Monaghan Morris/ Tim Soar
125–6	Photo: © Allford Hall Monaghan Morris/ Tim Soar
129	Photo: © Allford Hall Monaghan Morris/ Matt Chisnall
131	Image: Fiona McLachlan based on information supplied by Allford Hall Monaghan Morris
132	(top) Original image by Johannes Itten, redrawn by Fiona McLachlan (2011)
132	(bottom) Original diagram by Faber Birren, redrawn by Fiona McLachlan (2011)
133	Image courtesy of OMA Architects
134	Original image by Johannes Itten, redrawn by Fiona McLachlan (2011)
136	Image: © Sauerbruch Hutton
141	Photo: Jan Bitter Fotografie © bitterbredt.de
144	Photo: Jan Bitter Fotografie © bitterbredt.de
146	Images: © Gerhard Richter P_902.034, P_902.037, P_902.040
151	Image: © Gerhard Richter
158	Ben van Berkel/UN Studio, Photo: © Michael Moran
163	Photo: ©Ros Kavanagh/VIEW
169	(top) Ben van Berkel/UN Studio, Photo: © Christian Richter
169	(bottom) Ben van Berkel/UN Studio, Photo: © Iwan Baan
173–5	Ben van Berkel/UN Studio, Photo: © Christian Richter
176	Image: Wikipedia commons
179	Diagram redrawn by Fiona McLachlan (2011) after *A Color Notation*, Albert Henry Munsell (1905)
180	Designed by Whitbread Wilkinson 2011, under license from PANTONE
184	© Josef Albers/Yale University Press, published in Albers, J., *Interaction of Colours*, New Haven and London: Yale University Press (1963)
187	Original image by Johannes Itten, redrawn by Fiona McLachlan (2011)
190	Original image The Walpamur Co. Ltd (1959), redrawn by Fiona McLachlan (2011)
192	Photos: © Akzonobel/ICI Paints, Advertising Agency

INDEX

Page numbers in italic indicate photographs / drawings

accuracy / precision 105, 182, 191
achromatopsia 151
Adelaide Wharf flats, London 123–4, 135
aesthetic judgement 50, 112, 147, 150
Agora Theatre, Lelystad, Netherlands 13, 159, 165–70, *166–7, 170*
AHMM Architects *12,* 118–35, *118, 120–1, 124–6, 129–31, 193*
Akzo-Nobel / ICI / Dulux 123, 180, *192*
Albers, Josef 5, 11, 87, 91, 128, 140, 151, 154–5, *184,* 189, 193
Alsop, Will 63, *63, 164*
ambiguity / doubt 12, 18, 97–8, 101, 117, 140, 147, 155, 170
An Gaelárus Cultural Centre for the Irish Language, Derry, Northern Ireland *48,* 58, 61
Angel office building, Islington, London 128
Aristotle 178
artists, influence of 64–9, 194
artist/architects collaborations 63–4, 71–81, 104, 128–9
authority 9, 69–70, 89, 117, 128, 147–8, 174, 186

Bach, Johann Sebastian 72, 152
Balabanoff, Doreen 90–1, *91*
Ball, Philip 19–20
Barking Central, London *118,* 120, 128
Barragán, Luis 59, 96, 140
Batchelor, David 1, 69, 93
Bauhaus, the 5, *52,* 53, 140, 164, 186, 219, 220
Behnisch, Günther 64, 73
Berlage, Hendrix Petrus 26, 145
Birren, Faber 32, 101, 132, *132,* 178
Blackwood Golf Centre, Ireland *56,* 57
Blanc, Charles 193
Bolles Wilson Architects *17,* 18, 138–9

Braham, William ix, 27
Brandhorst Museum, Munich *26,* 155–7, *155–6*
brandscapes 175, 221
British Standards 180
Broëlberg, apartments, Zurich 106, 113
Brookfield Community Centre, Tallaght, Dublin 162–3, *163*
Brunnenhof apartments, Zurich 110, *111*
Buddhism 87, 101
Burke, Edmund 3, 204

Camerana, Bendetto 77–9, *77–8*
camouflage 11–12, 130, 145
Canter, David 121
Caruso St John Architects 29–47, *32–3, 35, 39, 41–5,* 64, 128, 192, 194
Castel Zuoz, Engadin, Switzerland 174–5, *174–5*
ceramic tiles *26,* 126–7, 145, 154–6, 182
chance/randomness in design 20, 69, 146–9, 157
Chapel of St Ignatius, Seattle, USA 84–5, *84–5*
character 42, 46, 50, 60, 101, 127
Chevreul, Michel Eugène 10, *31,* 31, 150, 179, 188, 219
Chipperfield, David (Architects) 95
Chmielewska, Ella 122
Chuco Reyes (Jésus Reyes Ferreira) 96
CMYK Co-ordinates 180–1, 223
Co-op Himmelblau 164, 220
Coates, Nigel 138
Cologne Cathedral, Germany 147
colour
    analogous colour 105, 107, 191
    atlas 178, 181
    balance 60, 130, 132–3, 149, 155, 157, 190
    blindness 152, 181, 205
    branding/identity 36–8, 117, 125–7, 130, 145, 163

colour (*continued*)
  brightness/intensity 105, 130–4, 149, 155
  choice 5, 177, 180–1, 191
  chronology 5, 38
  and contingency 70, 175
  contrast 10, 61, 74, 80, 132, 179–80, 191
  cultural tradition 29–30, 96–7, 101, 122, 194
  dynamic shifts 91–2, 191
  as an element of construction 14, 140
  equilibrium 130, 134, 137, 150–7
  experimentation 50, 58, 105, 112, 164–5, 186, 189
  forecasting 37–8
  harmony 60, 132–4, 150–7, 178, 189–91, 217
  immersive *3*, 89, 101, 112, 161, 168
  instability 134, 143, 145, 157
  as instrument 83, 88, 159, 161–3, 166, 174
  key 150–1, 207
  laws 10, 150, 152, 185–93, 217
  and light 10, 83–92, 191
  luminosity 20, 130–4
  as mask 5, 14, 36, 168–70
  matching 11, 182
  as material 53, 140
  metaphysical properties 3, 83–7, 97
  music, analogy with 72, 87, 100, 150–2, 178, 220
  navigation 176–83
  and personality 2, 192
  prejudice 1–2, 69, 73, 138, 185
  psycho-physiological effect 101, 141
  psychology 38, 121–2
  risk 18, 69, 130
  rules, and composition 7, 69, 147–9, 152–4, 185–93
  in school buildings 119, 125–7, 180, 189–90
  and the senses 87, 141–2, 175
  spectrum 10, 97, 150, 178
  systems 176–83
  synergies and discords *see also* dissonance 134, 137, 150–2
  temporality 87–92
  theory 1–2, 31, 71, 135, 157, 164, 177, 185–93

  as tool 3, 108, 140
  transformational qualities of 83–9, 159, 170–2
  tuning space with 2, 83, 162–3
  unreliability of 9, 69, 101, 161
  and the urban realm 119–27
  visual effects 54, 66, 137, 150, 161, 178, 190
  watercolour paintings 51, 53, *56*, *85*, 86, 96, 175
  wavelength 87, 97, 151
  wheels, spheres and solids 10, 176–9
'Colour Field' painters 67
Commission Internationale de L'eclairage (CIE color space) 179, 182
communication 177–83
complementary colours 130, 157, 188, 191
composition 67, 77, 145–8, 151–2, 190–1, 224
Constable, John 20
context *see also* site *and* place 7, 9
contiguous colours 7, 69, 155–7, 188, 190
contrast, simultaneous 5, 10, 69, 135, 154–5
contrast, successive 5, 11, 155
Cranbrook Institute for Science, Michigan, USA *88*, 87–9
Critical Regionalism 51, 92– 7
Crystal Palace/Great Exhibition building 31–4, 143
Cuisenaire Rods 100

decision-making 9, 147, 193–4, 205
decoration, *see also* surface 14–16, 34, 45–6, 113, 122, 138
Delacroix, Eugène 1
Deleuze, Gilles 175
Demand, Thomas 29, 42, *42*
Diener + Diener 79, 108
Diggelmannstrasse apartments, Zurich *93, 102, 109*, 110
digital colour/software 7, 182–3, 191
dissonance *see also* colour discords 134, 149–54, 190
Domenig, Günther 139
Donation Albers-Honegger, France 105–6, *106*
drawings, use of colour in 2, 51, 96, 138–40, 164, 221
dressing, of facades 14, 36, 40–7, 174
Duncan Hall, Rice University, Texas, USA 122

Eisenman, Peter 13
El Lissitzky 138–9, 160
emotion 1–3, 9, 40–6, 65, 69–71, 81, 89, 101, 123, 140–1, 159, 162, 189, 193–4
English landscape gardens 139
equilibrium 130, 134, 137, 150–7, 192
Erasmus Bridge, Rotterdam, Netherlands 162, 170
Ernst, Johannes 70–9
experimentation 50, 57, 79, 105, 112, 164–5, 186, 193

façade design 14, 18, 33–4, 36–45, 71–2, 98–9, 128, 145–57, 172–3
Falkenberg housing, Berlin 65
fashion 7, 34–9, 174, 188
Federle, Helmut 80
Ferrand, Pierre-Henri 117
Finlay, Victoria 19, 32
Fire and Police Station, Berlin *152, 153,* 152–3
Firminy, Church of St Pierre, France 84
Foreign Office Architects 108, 117
form, architectural 11–12, 51–4, 80, 142
Fra Angelico 51, *51,* 59
Frampton, Kenneth 92

Gage, John 101
Galbally, Social Housing, Ireland 13, 54, *54*
Garfield, Simon 34
gender 1, 5, 191, 193, 205
German Federal Environmental Agency, Dessau 140–2, *141*
Giedeon, Sigfried 161–62
Gigon/Guyer 16, 64, *93,* 102–17, *102, 105–7, 109, 111, 113–6,* 171, 192
glass, back-painted 152–3
Glucksman Gallery, Cork, Ireland 50, 57
Goethe, Johan Wolfgang von 5, 9, 10, 130–2, 150–1, 177–8
Gomez-Perez, Alberto 87, 139
Gothic and Classical 34, 45–6
Gotz, Lothar 43

Graves, Michael 18
Graz Music Theatre, Austria 168–9, *169*

Hadid, Zaha 13, 138–9, 160, *160*
Hamm, Oliver 74–5
Harbissson, Neil 151
Hård, Dr Andreas 179
Hassett, Grianne (Hassett Ducatez Architects) 162–4, *163*
Haus der Farbe, Zurich 93, *93,* 101, *195*
Hay, David 150
Heaney, Seamus 49, 212
Herbert, Laurence 180
Hering, Ewald 179, 191
Herzog & de Meuron 18, 80, 107–8, 117, 211
Hilmer & Sattler 79
Hogarth, William 20
Holl, Steven (Architects) 4, 53, *82, 84, 85, 88, 90, 97, 98–9,* 83–101, 139, 194
Howth House, County Dublin 56–7
hue, value, chroma 178–82, *179,* 189–90, 223
Hutton, Graeme 103

illusion, colour as intellectual 5, 12–14, 84–7, 143, 165–8, 188
Impressionists, the 10, 19–20, 145
indexing, of colour 7, 177–83
Ingle, Charlotte 64, 128, *129*
interaction, of colours 5, 7, 11, *184,* 188–93
International Color Consortium (ICC) 181
intuition 37, 50, 63, 67, 150, 164, 186
Irish Film Centre, Dublin 23, 58, *59,* 61
Irish Pavilion, Museum of Modern Art, Dublin *55,* 56
iron oxide 22–3, 104, 208
irrationality 9, 194
irregularity 134, 145–7, 178
Itten, Johannes 5, 10–11, *52,* 53, 130, 132, *134,* 157, 164, 178, 185–6, 192

Jarman, Derek 9, 123
Jones, Owen 29–31, 34, 143
Judd, Donald 67, *67,* 77, 120, 162–4

Judge Institute, Cambridge, England 122, 122

Kandinsky, Wassily 73, 119, 150, 200
Kapoor, Anish 58
Kay, Paul and Berlin, Brent 123
Kentish Town Hall, London 121, *121*, 128
Kix Bar, Vienna, Austria 13, *13*
Klee, Paul 74
Klein, Yves 20, 74, 214
Knut Hamsun Museum, Norway 89–90, *90*
Koolhaas, Rem/OMA 18, 38, 107, 133, 138, 170
Krischanitz, Adolf 8, *79*, 79–81
Kroll, Lucien 72, 145

La Defense Offices, Almere, Netherlands 170–4, *171, 172*, 175
La Tourette, Chapel at Monastery 84, 151
Laban Dance Centre, London 80
Lancaster, Michael 186
language, of colour 123, 177–83, 188
Laycock, Kevin 150
Le Corbusier 12, *22*, 54, *54*, 57, 64–7, 74, 81, 84, 100, 139–40, 147–8, 152–4, *154*, 160–1, 182, 186
Leatherbarrow and Mostafavi 14, 106–7, 145
Léger, Fernand 117, 213
Legorreta +Legorreta *94*, 95–6, 101
Lenclos, Jean-Phillipe (and Lenclos, Dominique) 92–3, 128, 142
Libeskind, Daniel 139
light
    coloured 10, 86–91, 173–5
    conditions 11, 16, 45, 69, 85, 89–91, 95, 112, 170, 172, 191
light reflectance value (LRV) 180, 189
lime / limewash 19, 22–4, 57, 61, 208
Linked Hybrid apartments, Beijing, China 100–1
Longyearbyen, Norway 50, 101, 216
Loos, Adolf 14–16, 30, 36, 40–2, 130
Lootsma, Bart 164
Lynch, Kevin 121
Malevich, Kasimir 139, 160

Matisse, Henri 69
MAXXI Museum of Twenty-First Century Art, Rome *160,* 160
McGuire, Brian 55–6
McLachlan, E & F, Architects *5, 10*
McLaughlin, Niall, Architects 27, 209, 222
McLean, Bruce 63
meaning – and association 2, 7, 18, 42, 55–6, 61, 63, 108, 115–17, 120–5, 147, 178, 207
memory 55, 91, 93, 101, 104, 122, 125
metamerism 181, 191, 215
metaphysics 87, 100
models, use of 51, 76, 84, 115, 127, 157, 163–4
'Modulor', the 151, 201
Mondrian, Piet 74
Monet, Claude 10, 112, 188
Monsoon Headquarters, London 125, 130
Moretti, Simon 34
Mozart, Amadeus 186–7
Mueller & Sigrist 95
Muir, Jean 36–7
Müller House, Prague 40–2
Müller, Harald, F. 64, 104, 106, 117
Munsell, Albert 178, *179*
Murcia Town Hall, Spain 151

Natural Color System (NCS) *37*, 38, 50, 93, 101, 149, 178–9, 191, 208
Newman, Barnett 3, 67, 69
Newton, Issac 9, 10, 132, 150, 178, 220, 222
nomenclature, of colour 2, 19, 123
Nordensen, Guy (Engineers) 98–100
Nottingham Contemporary Gallery, Nottingham, England 29, 30, 43–6, *43, 44, 45, 46, 47*

O'Donnell + Tuomey 13, *23*, 25, 25, 48–61, *48, 53–6, 59–60*, 73, 130, 163, 192, 194
O'Toole, Shane 4
Oberascher, Leonhard 91–2, *92*
objective principles 5, 9, 185–6, 189–91
Onkel Tom's Hütte 65, 65–7, 80

Opponent Colour Theory 179, 191
optical effects 84, 110, 140, 143, 154–7, 163, 188, 190
ornament 16, 30, 32, 45, 140
Ostwald, William 152, 206
Outram, John 16, *122*, 122–3, 189
Ozenfant, Amédée 64, 81, 152, 206

paint 9–10, 14–16, 18, *21*, 21–6, 43–6, 74, 112–16, 122–4, 155, 163, 170, 177, 180–2, 191
palette, colour 4–7, 34, *37*, 49–61, 92–5, 101, 125–31, *131*, 147, 154, 180, 192
Pallasmaa, Juhani 103
Pantone 180–1, *180*, 182
Passini, Romedi 127
pattern 11, 14, 16–21, 29–34, 40, 45–6, 71, 87, 122, 143–9, 154
perception 4, 7, 10–12, 67–70, 83–5, 87, 89–91, 101, 130, 140–1, 155–7, 161–2, 170, 178, 190–3
Perkin, William 34
Pessac, France, housing at 54, 66
Pflegi-Areal apartments, Zurich, Switzerland 115, *115, 116*
phenomenology 101, 103, 116, 141, 190
physiology 178
Pichler, Walter 107
Picturesque, the 29, 141, 145
pigments
    historical development of 18–20, 30–1, 51, 58, 92–3, 210
    technology of 18–26, 29, 34, 37, 103, 142, 208
Pilotengasse housing, Vienna, Austria *79*, 79–80
pixelation 71, 145
place, sense of – *see also* context 46, 50, 91–5, 101, 103–4, 117, 139, 141–2
pointillism 137, 154–5
Pollock, Jackson 20
polychromy 4, 12–13, 32–4, 43, 64, 66, 101, 104, 122, 129, 137–57, 148
pragmatism 50, 135, 189, 193–4
Pugin, Augustus Welby 14, *28*, 29, 34, 45
Putz, Oscar 13, *13*, 79–81

Radiant Colour/Light Film (3M) 27, 171–4, 208, 221
Raines Court flats, London 129–30
RAL colour system 61, 134, 179, 212
Rama, Eddi 18
Ranelagh Primary School, Dublin, Ireland 54, 58–60, *60*
rationality 9, 89, 193
Rattray, Charles 103
Raumplan 40
regulating lines 100, 145–8, 151, 186, 225
Reitveld, Gerrit – Schröder House 12, 16, 74
Rendell, Jane 117
representation, of colour 2, 134, 139, 176–9, 182–3
rhythm 69, 72, 76, 100, 147–9, 151–2, 157
Richter, Gerard 16, *146*, 147, *151*, 151
Riley, Bridget *68*, 69, 73, 162, 165, 186
Rossi, Aldo 2, 139
Rothko, Mark 3, 16, 20, 67, 70, 218
Rüegg, Arthur 66–7, 154
Runge, Phillip Otto 5, 132, 152, *176*, 176–9, 187
Ruskin, John 70

Salter, Peter 138
Salubra – Le Corbusier 22, *22*, 57, 154, 182, 213
Salvation Army building, Paris 154
Sarphatistraat, Amsterdam *82*, 89, 97, *97*, 100
Sauerbruch Hutton Architects 12, 25, *26*, 136–57, *136, 141–4, 148–9, 152–3, 155–6*, 182, 193–4
Scharoun, Hans 45
Schiess, Adrian 16, 110–17, 171
Schinkel, Karl Friedrich 30, 63
Schumacher, Patrik 160
Sean O'Casey Community Centre, Dublin, Ireland 25, *25*, 53–4, *53*, 61
Sedus warehouse, Germany *144*, 145, 157
Semper, Gottfried 14, 30, 42–3, 113
seriality 57, 130
Shaw, D. E offices, New York, USA 83–4
signing/signage 120–2
Simmons Hall of Residence, MIT, Boston, USA *98–9*, 100
simultaneous contrast 10, 69, 137, 154–5, 188

Index 233

site, influence of 30, 50–1, 61, 91–7, 103–5, 115, 125–6, 140, 142, 181–2, 191
Smedal, Grete 50, 101
Soane, Sir John 29
space and time 160–2
spatial
    composition 2, 5, 12–14, 42–3, 46, 57, 116, 119, 134, 139–40, 157, 160–1, 163–4
    effects 57, 65–6, 81, 101, 189, 190, 193
    energy 89
spectrophotometric curve 181
Sports Centre, Davos, Switzerland 104
standard light conditions 183
standardization, international / ISO 181
Star Palace, Kaohsiung, Taiwan *173*, 173–4
Steidle Architekten/Otto Steidle *62, 63*, 70–81, *72, 74–6*, 128
Stellwerk, Zurich, Switzerland 104–5, *105*, 113
Stirling Castle, Scotland 24
Stirling, James (Michael Wilford) 2, 49, *49*
Studio Myerscough / Morag Myerscough 120–1, 127, 129
Stuttgart *Weissenhofsiedlung*, Germany 66
subjectivity 5, 7, 50, 71, 89, 147, 175, 185–93
Sublime, the 3–4, 193
Sullivan and Adler 30, 43, 45
supergraphics 104, 120–6
surface 9–27, 30, 36, 40–5, 71, 76, 115–17, 138–45, 154–7, 159–61, 164, 168, 189
surface
    scripting 122
    reflective 16, 110–12
Susenbergstrasse apartments, Zurich, Switzerland *114*, 115
sustainability 141–2, 149–50
symbolism 16, 83, 122, 189
Syme, Patrick and Werner, Abraham Gottlob 123
synaesthesia 150
Szyszkowitz + Kowalski *95*, 95–6, *96*, 139

Taut, Bruno 64–7, *65*, 72, 74, 80
Templar House flats, Harrow, England 128–9, *129, 130*

Tirana, Albania *17*, 18
Titian 20, 208
tolerance 182, 191
Turin, Olympic village, Italy 77–9, *77, 78*
Turrell, James 3–4, *3*, 140, 175

Ulm University, Germany 72–3, *72*
ultramarine 19, 22–3, 34, 74
UN Studio 13, *15, 158*, 158–75, *166–7, 169–75*, 194
Unity Tower, Liverpool, England 128
urban colour 18, 66, 74–7, 79–80, 119–30, 142, 148, 186

van Doesburg, Theo 12, 66
van Eyck, Aldo 186
van Gogh, Vincent 10
Venturi, Robert 18, 150, 154
Victoria & Albert Museum of Childhood, London 29, *32, 33*
Villa am Römerholz, Winterthur, Switzerland 104–5, 107–8
Villa La Roche, Paris, France 12, 160
Villaverde housing, Madrid, Spain 95
visual balance 130, 149, 154
visual stimulation 141
Vreisendorp, Madelon 138

Walsall, New Art Gallery, England *39*, 39–46, *41*
Wang, Wilfred 107–8
watercolour painting 20, 51, 56, 84, *85*, 175
Webb, Philip 145
well-being 90, 137, 140–2, 189
West 8, 165
Westminster Academy, London 119, 123–8
Whittle, Steven 150
Whyte, Iain Boyd 64
Wiesner, Eric 64, 70–80, 192
Wigley, Mark 36, 64, 139
Wittgenstein, Ludwig 187, 193
wrapping, of facades 12, 40–3, 145

Zenghelis, Elia 138
Zumthor, Peter 108